Extreme Science
Chasing the Ghost Bat

Extreme Science

Chasing the Ghost Bat
and Other Mysteries of Nature

The Editors of **SCIENTIFIC AMERICAN**

Peter Jedicke, Project Editor

ST. MARTIN'S GRIFFIN

NEW YORK

A Byron Preiss Book

A Byron Preiss Book.
Project Editor: Peter Jedicke
Project Coordinator: Howard Zimmerman
Assistant Editor: Carlos Hiraldo
Design by Tom Draper Design
www. stmartins.com

The book's essays first appeared in the pages of *Scientific American Magazine* as follows: "The Man-Apes of South Africa," May 1948; "The Hominids of East Turkana," August 1978; "The Omnivorous Chimpanzee," January 1973; "On the Origin of Subspecies," March 1999; "Survivors and a Winner," May 1999; "Killer Kangaroos and Other Murderous Marsupials," May 1999; "The Terror Birds of South America," February 1994; "Shrews," August 1954; "Snakebite," January 1957; "Chasing the Ghost Bat," June 1999; "Running on Water," September 1997; "Diving Adaptations of the Weddel Seal," June 1987; "Stalking the Wild Dugong," September 1998; "Secrets of the Slime Hag," October 1998; "The Giant Squid," April 1982; The Army Ant," June 1948; "Slavery in Ants," June 1975; "The Fire Ant," March 1958; "Singing Caterpillars, Ants and Symbiosis" October 1992; "Mating Strategies of Spiders," November 1998; "Africanized Bees in the U.S.," December 1993; "Predatory Wasps," April 1963; "Plants that Warm Themselves," March 1997; "The Voodoo Lily," July 1966; "Carnivorous Plants," February 1978; "The Toxins of Cyanobacteria," January 1994; "The Lurking Perils of *Pfiesteria*," August 1999; "How Killer Cells Kill," January 1988; "Predatory Fungi," July 1958; "Extremophiles," April 1997.

Library of Congress Cataloging-in-Pulication Data

Extreme science: Chasing the ghost bat and other mysteries of nature / the editors of
Scientific American; Peter Jedicke, project editor.
    p.   cm.—(Extreme science)
Includes bibliographical references and index.
"A Byron Preiss Book."
ISBN: 0-312- 26818-1
1. Biology—Miscellanea.   I. Jedicke, Peter.   II. Scientific American.   III. Series.

QH349.C43   2001
570—dc21

                                                                              2001042406

First Edition: November 2001

10   9   8   7   6   5   4   3   2   1

CONTENTS

Preface

A persistent theme in popular fiction is life on other planets, and a myriad of odd forms have been invented in print and on film. Do you remember the eight-legged beasts of burden on Edgar Rice Burroughs's Mars, the thoats? Or Larry Niven's Kzinti? I always thought of the Kzinti whenever I beheld Chewbacca in *Star Wars*. From the menagerie of critters in Hollywood's *Men in Black* to the kindness of the Mother Thing in Robert A. Heinlein's *Have Space Suit—Will Travel*, writers have come up with an amazing variety of imaginary life-forms.

And yet, fantasy can't touch nature. The real world of biology and zoology presents us with more than enough wonder, and plenty of mysteries, to occupy our wildest dreams—or nightmares. The articles in this book, selected from the pages of *Scientific American*, make that point vividly. The tree of life—a conceptual scheme for arranging all the species—begins with forms so simple that an individual consists of barely a single cell, and branches again and again so that shapes as weirdly different as the northern ghost bat and the giant squid (see Chapter 3) are all encompassed. These are the living things that share this planet, our planet, with us. The same principles of life that determine our own existence also shape these strange rival forms. Our genetic material is made of the same basic chemicals as theirs. And perhaps most significantly, our past is entwined with theirs. To say nothing of our future.

Our survey begins with humankind—ourselves. The controversial search for our origins has gone on for well over a century. Humans have found our place on the tree of life, and we found it in Africa (see Chapter 1). Perhaps it is only because of the perpetual objections of many religiously minded folks that the scientifically educated still need to learn about Charles Darwin and the naturalists of the mid-19th century. When Darwin finally published *The Origin of Species* in 1859, the theory of evolution did not spring solely from his imagination. During a five-year voyage aboard the H.M.S. *Beagle* in the southern hemisphere, Darwin collected and observed hundreds of species. He sent specimens and notes back to England, and some of what he wrote was published privately even before he returned. Being by nature a methodical researcher and not inclined to engage dispute, Darwin was in no hurry to formulate a theory.

Classification of living things into a tree structure was well established by Darwin's time, and there had been talk of some kind of evolution for decades before. Then, as now, the issue centered on the critical question: is it possible for new species to develop, or are the species currently present all there ever has been since the world began? Baron Georges Cuvier had developed a skill for looking beneath the surfaces of animals and explaining relationships among the skeletal structures of different species on the basis of the animals' behavior and abilities. He also studied fossil skeletons of species obviously extinct, such as the pterodactyl, which was exhibited in 1812, and he nonetheless found a place for them on the tree of life. Motivated by devotion to the literal truth of the Bible, Cuvier devised a scheme of repeated worldwide floods that caused the extinction of some species and cleared the way for new.

Then, as now, the issue was much more complicated than just a single central question. If new species could develop, did they appear suddenly or gradually? Were new species related to preexisting ones, perhaps by being progressive improvements? Another gentleman scientist, like Darwin and Cuvier, was Charles Lyell. Lyell started with the view, not uncommon in the early 19th century, that subsequent generations might inherit characteristics solely because those characteristics were used or developed by the current generation of living things. The famous example was the giraffe: each generation was thought to have lengthened their necks by stretching for leaves on higher limbs of trees, and to have passed the gradually lengthened neck on to the next generation. But Lyell eventually recognized that even internal organs needed to be part of the scheme, and it was just too difficult to believe that creatures could "stretch" their internal organs. So Lyell gave up—temporarily—on the idea of a gradual change in the nature of species and tried instead to devise a theory that would explain the observed distribution of species from one continent to another.

While Darwin and the *Beagle* were relaxing in Montevideo in the fall of 1832, a copy of Lyell's book *Principles of Geology* arrived. Lyell's thorough analysis showed that similar geographic regions in different parts of the world mostly had distinctive flora and fauna. Lyell suggested this was peculiar because species were thought to be specialized for their conditions. Then why would variety exist in similar circumstances? A slow dispersion of many species out of some starting point and into differing harsh conditions might explain how some species could become extinct in certain areas but not others. But could all species have started together? Could all the variety that we see today be what is left after expansion from a single beginning? Lyell suspected that new species might be created or somehow come into being on a regular basis, but he had no details to offer as to how this could happen.

All of this, from Lyell's book, was on Darwin's mind as the *Beagle* continued down the coast of Argentina and on to the Galapagos Islands. There, Darwin was impressed by both the diversity and the similarity of species like the many finches and the giant tortoise. Continuing to work with his notes and specimens after his return to England, Darwin wove another concept into his work: population pressure. The numbers of any species in a favorable situation would grow and grow until the impact of the species on the environment was sufficient to reduce or to eliminate the advantages of the situation. This concept was articulated by Thomas Malthus, and it was not a popular idea because Malthus extended its application even to the human race. Darwin was more interested in what the idea added to his developing theory of species: competition for resources was a selective agent.

Ever cautious, Darwin worked for many years on an introductory paper that, in the end, was not even published in his lifetime. But the theory was essentially complete. Species changed gradually because the pressure of cruel nature eliminated unsuitable individuals before they could reproduce. Darwin did not make any conclusions about what this theory might do for religion. Instead, he focused on the shortcomings that any scientific theory has in its early days. He felt more observations were needed, despite his thoroughness. But, most important, he knew that he was still missing an explanation for how the variation of characteristics was passed from one generation to the next. The vagaries of science worked against him on this one point, because the secret of heredity was discovered by Gregor Mendel not long after Darwin's theory came out. But Mendel's work, although published, was not widely read, and it lay unknown until decades later, after both Darwin and Mendel were dead.

Another naturalist, Alfred Wallace, helped stir Darwin to stop laboring on his observations and finally to announce his theory. Wallace was at work in southeast Asia, and his own observations about species there brought him to much the same conclusions as Darwin. Darwin was famous enough that Wallace knew of him, and Wallace sent a manuscript describing his work to Darwin even though Wallace did not know specifically that Darwin had a similar theory. Encouraged by Lyell, Darwin dashed off a manuscript of his own and presented both his work and Wallace's together to the scientific community.

The publication of *The Origin of Species* a year later was the beginning rather than the end of the story. Scientists and scholars everywhere lined up for or against Darwin's theory. It was immediately obvious that the theory could be applied to humankind as well as to finches. Wallace and others even connected the theory to social development. But it was biological evolution of humankind that stirred up a controversy that will not rest. Thomas Huxley was another naturalist who entered the fray after reading *The Origin of Species*. Not so withdrawn as Darwin, Huxley became the front for the theory of evolution. Just seven months after *The Origin of Species* appeared, at a meeting of the British Association for the Advancement of Science, Huxley had a chance to stand up for evolution against Samuel Wilberforce, the bishop of Oxford, a respected spokesperson for religious views. The incident has become one of the most famous in all of science. The bishop ended his speech by asking wittily if Huxley was descended from apes on his father's side or his mother's side. Huxley then ended his speech by saying that he would rather have an ape as his ancestor than a learned man who would employ ridicule in a serious debate.

Darwin eventually laid down the gauntlet in another book, *The Descent of Man*. Although there are many religously devoted persons who still today hold that Darwin's view of evolution is not correct, time and the weight of scientific evidence have really settled the issue. Evolution has the status of a law in science, like gravity or relativity, and it is taken for granted in the unfolding of the tree of life. The variety of ants (see Chapter 4) and bees and wasps (see Chapter 5) that abound in our world are explained by evolutionary developments, and any issues that arise from modern biology must take evolution into account.

Plants (see Chapter 6) are equal members of the evolutionary tree of life, even if members of the animal kingdom get most of the attention. When Mendel proved that the mechanism of heredity could pass on certain characteristics, it was plants that he was studying. Finally, our survey of the wild and wonderful tree of life concludes with some minuscule examples (see Chapter 7) of just how strange things truly are on this planet that we call home. There's plenty of what makes science fiction so engrossing right here, right on our own tree of life.

—*Peter Jedicke*

Chapter 1

The Most Dangerous Animal of All

The Man-Apes of South Africa

Wilton M. Krogman

Man is known to have existed on this planet for at least one million years. He can trace back his line with fair confidence to *Pithecanthropus* and *Sinanthropus*, the Java and China fossil men, who from the evidence of bones are the earliest known creatures that may be considered truly human.

Superficial resemblance has argued since Darwin that man is an offshoot from the anthropoid apes. But the case remains unproved; indeed, the more it is examined, the less convincing it becomes. Between the most highly developed modern anthropoids and the most primitive living men exist differences so basic that the two can be connected only by imagining a series of "missing links" of an unimaginably paradoxical pattern—here progressing, there retrogressing, and in some characteristics suddenly striking off in an unpredictable new direction.

The search for evolutionary links between man and his elusive ancestors, whoever they may have been, is one of the most fascinating in all science. It has engaged the unremitting attention of fossil hunters for a period of more than a century. In South Africa, Dr. Robert Broom discovered fossil bones that compelled us to recast our thinking as to when and where man made his first appearance on the Earth. The bones make up a number of subhuman skeletons. These South African "Man-Apes," of which the best-known is named *Plesianthropus* (from the Greek *plesio*, meaning "close to," and *anthropos*, "man"), are truly astonishing connecting—not missing!—links between man and the apelike forms from which he arose. Notice that the word is "apelike." If we adhere to the term "apelike" or "anthropoid" in describing man's ancestors, we shall avoid prejudging the question of man's exact relationship to the apes.

The Plesianthropus story begins with a dramatic prologue. Dr. Raymond Dart, professor of anatomy at the University of Witwatersrand, was keeping a sharp eye on the digging in an old limestone quarry near Taungs, in Bechuanaland, where many fossil apes had been found. There, from the solid rock, was dug one day a limestone cast of the inside of a skull. Its shape was so unusual that Professor Dart pressed on; soon his diggers unearthed the front half of a skull top and an almost complete facial skeleton, all clearly belonging to the same individual. The skull had a full set of deciduous, or baby, teeth, plus the first permanent molar. The fossil was therefore that of a youngster about six human years old.

Some parts of this skeleton were definitely apelike. But others, especially the teeth, were very much like man's. Indeed, if the teeth had been found separately, they would have been pronounced human. The whole skull showed such an amazing blend of anthropoid and human traits that Dart immediately concluded that he had found an important intermediate form. He gave it the name *Australopithecus africanus* (meaning southern ape of Africa).

Dart's find was fascinating but inconclusive. It is hard to be sure about an immature skeleton: the skulls of a young anthropoid and a young child are much more alike than those of adults.

Then, in 1936, Dr. Broom, chief paleontologist of the Transvaal Museum in Pretoria, found several fossil fragments that fitted together to form part of a skull. This skull was adult. Broom first called his find *Plesianthropus transvaalensis*.

Two years later at Kromdraai, just two miles away, a schoolboy casually picked at a weathered outcrop of fossil-bearing limestone. He stuffed some bits of bone and teeth into his pockets, along with other boyish treasures, and went his way. When Broom accidentally heard of these acquisitions, he tracked the boy down. In his trouser pocket were four priceless teeth, the remains of still another member of the same subfamily as Australopithecus and Plesianthropus. Broom called this third genus *Paranthropus* (meaning manlike) *robustus*.

The hunt grew vigorous. It was interrupted by the war, but was resumed with immediate and brilliant success. Broom found a number of excellently preserved fossils, including a complete skull and an "almost perfect pelvis, femur, tibia, a number of ribs, some vertebrae and a crushed skull" of an adult female about four feet tall.

Plesianthropus is represented literally from head to toe, Paranthropus is represented by most of the palate, the left side of the face with the cheek-bone arch, the right side of the lower jaw with most of the teeth, parts of the skull-base, the skull side-wall and fragments of arm, hand and foot bones.

Here, then, is an extraordinarily complete collection of bones, outlining a group of creatures that stand somewhere between ancient anthropoids and man. Now, the first thing the anthropologist wants to know about these bones is their age. To fit them into a family tree, we must place them in time with relation to the other anthropoid (apelike) and hominid (man-like) fossil remains.

The four most recent geological periods, counting back from the present, are: the Pleistocene (the last million years), the Pliocene (the preceding six million years), the Miocene (the 12 million years before that) and the Oligocene (the 16 million years before that). The earliest men, Pithecanthropus and Sinanthropus, are generally placed in the early Pleistocene. The earliest known anthropoid forms go back to the Oligocene.

The evidence of plant and animal fossils with which Broom's and Dart's finds were associated appears to put them in the Pliocene period. Plesianthropus's bones, for example, were found associated with remains of a fossil baboon, *Parapapio*. Parapapio, in turn, was linked with a fossil hyena whose name is *Lycyaena*. And Lycyaena is definitely dated in the early Pliocene.

If we accept the paleontological evidence we may estimate that the South African Man-Apes lived about seven million years ago.

Now let us look at the bones themselves. The three South African genera—Australopithecus, Plesianthropus and Paranthropus—are grouped as the subfamily *Australopithecinae*, belonging to the family *Hominidae*. Because of their skeletal similarities, we can consider all three genera together.

We start with the size of the brain, the most important single measure in distinguishing man from the other primates. The Australopithecines had a small skull, closer to an ape's than a man's. The brain volume of Plesianthropus ranged from about 400 to 700 cubic centimeters, that of Paranthropus was probably from 440 to 750, and Australopithecus' was from 500 to 800.

This is an improvement on the apes. Their brain sizes range roughly from 300 to 600 cc; the largest ape brain ever found, that of a male gorilla, was 620 cc. But in brain size Plesianthropus and his fellows were definitely subhuman. The most primitive true man, Pithecanthropus, had a brain of 650 to 1,000 cc. Neanderthal man, who lived about 50,000 years ago, ranged from 1,070 to 1,730 cc. Modern man, excluding pathological cases, varies in brain volume from 860 to 2,000 cc.

It seems clear, then, that the Australopithecines had slightly larger brains than anthropoid apes and overlapped into the lower range of the early hominids. But they did not quite reach the lowest limit of *Homo sapiens*.

The next important clue is the teeth. Because they are the hardest part of the skeleton and hence the last to disintegrate, teeth constitute the principal remains of our paleontological past. Without going into precise dental detail, it can be said that the Australopithecines' teeth are unmistakably manlike. The structure and form of the deciduous and permanent teeth, their eruption time and order and their alignment in the jawbones are all human. Moreover, the pattern of toothwear and the way the jawbone is hinged to the temporal bone at the base of the skull suggest that Plesianthropus chewed his food with a rotary grinding motion, like man, instead of chomping it, like the anthropoids.

What did Plesianthropus look like? His facial profile was relatively manlike; it was losing the anthropoid slope. The bony ridges above his eye sockets were less protruding, and he had the beginnings of an arching human forehead. Also like man, he had fairly slender cheekbone arches. The front surfaces of his upper jaw were vertical, not slanting as in apes. The "simian shelf," a characteristically anthropoid platelet of bone on the underside of the lower jaw, had all but disappeared. And on the front of the lower jaw was a little knob—the beginning of a human chin.

Plesianthropus's bones retained many anthropoid features. The side teeth in his upper jawbone, for instance, were arranged like an ape's. The contactline, or "suture," between the two bones of the hard palate, which in man usually closes at birth, appears to have delayed in closing, as with an ape.

Below the neck, Plesianthropus looked like an incompletely developed biped. His elbow joint did not flex as freely as man's; his arms seem to have been permanently bent. His anklebone, intermediate between a human and an apelike form, indicates that the ankle may have been designed for weight bearing, as in an erect man, but that the foot may still have possessed a grasping big toe. The bone in the center of his palm suggests a hand freed from locomotion but with limited flexibility for manipulation. His thighbone and hipbone, in the opinion of some, were mechanically adapted for standing, walking and running in the erect position.

Did Plesianthropus actually walk upright? All the evidence points to the likelihood that he did. One significant sign is the position of the foramen magnum and the occipital condyles. These structures form the joint where the skull rests on the backbone. In anthropoids, this joint is far back on the skull-base. But in man, it is in the middle, so that his skull is balanced squarely on the spinal column. Plesianthropus was well-advanced toward this balanced position. His foramen magnum was much farther forward than that of the anthropoids, but not as far forward as man's.

As to the quality of Plesianthropus's brain and his abilities, we have little evidence. The casts of the inside of his skull do not show his brain configurations in any accurate detail. They tell us only the approximate size of regions in his brain.

In the view of the English anthropologist W. E. Le Gros Clark. The endocranial casts suggest "that the mental powers of the Australopithecines were probably not much superior to those of the chimpanzee and gorilla." Nonetheless, on the basis of another piece of evidence, Le Gros Clark concedes that their intelligence was "definitely in advance" of that possessed by the modern apes. In the same cave with some remains of Australopithecus were found the crushed skulls of baboons, suggesting that the mental powers of Australopithecus were great enough to enable him to hunt and kill monkeys.

What does our evidence on the Australopithecines add up to? It is this: that in the Pliocene period, about seven million years ago, there lived a form that was intermediate between anthropoid and man. He had a brain near the anthropoid, a dentition practically human and a general skeletal build well-adapted to the human upright position and locomotion. The Australopithecines fulfill almost every requirement of a real connecting form. Moreover, they show that the rate of evolution differs in various parts of the body: thus dentition is ahead of long bones, and long bones are ahead of brain. The evidence of Plesianthropus's teeth strongly suggests that he and his fellows may have been ancestors of the first true men, Pithecanthropus and Sinanthropus. (*Gigantopithecus* of Java may fit into the line here, but there is some doubt whether he was anthropoid or hominid.) From the Java and China men, in turn, via Solo man of Java and Neanderthal man of Europe, finally came Homo sapiens.

Where do the apes fit into this picture? There is increasing conviction that the modern anthropoids (gibbon, orangutan, chimpanzee and gorilla) have arisen independently of man. Somewhere in the Oligocene or the Miocene, their ancestors split off from the common trunk to form separate branches.

The Australopithecines may, indeed, have been an offshoot from some still undiscovered ancestral form. The modern apes and anthropoids may each have come from a different ancestor—and I suggest that human evolution is almost a straight-line parallel of anthropoid evolution, with a basic split off of the several lines in the Oligocene period 30 million years ago.

It is the South African Man-Apes who have pointed the way to this conclusion. In effect they have emancipated man from a presumed simian ancestry. "Family tree" may be a misnomer. Perhaps we never were brachiators, i.e., swingers through the trees by our arms. It is entirely possible that for millions of years we've walked erect, striding head up toward our evolutionary destiny, whatever it may be.

The Hominids of East Turkana

Alan Walker and Richard E. F. Leakey

The continent of Africa is rich in the fossilized remains of extinct mammals, and one of the richest repositories of such remains is located in Kenya, near the border with Ethiopia. The first European to explore the area was a 19th-century geographer, Count Samuel Teleki, who reached the forbidding eastern shore of an unmapped brackish lake there early in 1888. Exercising the explorer's prerogative, he named the 2,500-square-mile body of water after the Austro-Hungarian emperor Franz Josef's son and heir, the archduke Rudolf (who within a year had committed suicide in the notorious Mayerling episode). Teleki's party evidently passed a major landmark on the shore of the lake, the Koobi Fora promontory, during the last week in March. They took notes on the local geology and even collected a few fossil shells, but they missed the mammal remains.

Teleki's name has faded into history, and even the name of the lake has now been changed by the government of Kenya from Lake Rudolf to Lake Turkana. Yet today the Koobi Fora area is world-famous among students of early man. Its mammalian fossils include the partial remains of some 150 individual hominids, early relatives of modern man. They represent the most abundant and varied assemblage of early hominid fossils found so far anywhere in the world.

The fossil beds of East Turkana (formerly East Rudolf) might have been found at any time after Teleki's reconnaissance. It was not until 1967, however, that the deposits came to notice. At that time an international group was authorized by the government of Ethiopia to study the geology of a remote southern corner of the country: the valley of the Omo River, a tributary of Lake Turkana. Erosion in the area has exposed sedimentary strata extending backward in time from the Pleistocene to the Pliocene, that is, from about one million to about four million years ago.

Supplies going to the Omo camps were flown over the East Turkana area: on one such trip one of us (Leakey) noticed that part of the terrain consisted of sedimentary beds that had been dissected by streams and that appeared to be potentially fossil-bearing. A brief survey afterward by helicopter revealed that the exposed sediments contained not only mammalian fossils but also stone tools. This reconnaissance was followed up in 1968 by an expedition to the vast, hot and inhospitable area. Out of a total area of several thousand square kilometers the expedition located some 800 square kilometers of fossil-bearing sediments, mainly in the vicinity of Koobi Fora, Ileret and to the south at Allia Bay.

The large task of establishing the geological context of both the fossils and the stone tools began in 1969. The work that season was highlighted by the excavation of the first stone tools to be found in stratified sequences there and by the discovery of two skulls of early hominids. It was with these discoveries that the enormous importance of the area for the study of human evolution began to be recognized.

A formal organization was established: the Koobi Fora Research Project. The project operates under the joint leadership of one of us (Leakey) and Glynn Isaac of the University of California at Berkeley. In the years since it was founded the project has brought together workers from many countries who represent many different disciplines: geology, geophysics, paleontology, anatomy, archaeology, ecology. The interaction of specialists has become a particular strength of the project as the workers have become increasingly aware of the particular outlook (and also the limitations) of fields other than their own.

The project area extends from the Kenya-Ethiopia border on the north to a point south of Allia Bay where the land surface is of volcanic origin. The western boundary is the lakeshore and the eastern is marked by another volcanic outcropping. The promontory of Koobi Fora itself, a spit of land that extends a few hundred meters into the lake, is the site of our base camp. Each of the three principal areas of fossil-bearing sediments has its own natural boundaries; when these areas are seen from the air, they show up as pale patches among the darker volcanic terrain.

In studies such as these it is of paramount importance to develop a chronological framework that allows the fossil finds to be placed in their correct relative positions. The construction of such a framework is the responsibility of the project geologists and geophysicists. The geology of East Turkana is straightforward in its broad outlines but extremely complex in many of its local details. Among the factors responsible for its complexity are abrupt lateral shifts in the composition of the exposed sedimentary strata, discontinuities, faulting that involves rather small displacements and above all the absence from many of the sediments of volcanic tuff: layers of ash that play a key role in correlating the strata.

The difficult work of geological mapping and stratigraphic correlation has been carried out by many of our colleagues but mainly by Bruce Bowen working in collaboration with Ian Findlater and Carl F. Vondra. The complexities have nonetheless prevented the accurate placement of some of the most important early hominid fossils in the stratigraphic framework the geologists established. In such cases we have provisionally assigned the specimens to temporal positions on the basis of criteria other than stratigraphic ones.

The fossil-rich sediments are underlain by older rocks of volcanic origin. The sediments themselves are of various kinds, laid down in such different ancient environments as stream channels and their associated floodplains, lake bottoms, stream deltas and former lakeshores. For the most part the strata dip gently toward the present Lake Turkana. In the past, extensions of various deltas and coastal plains frequently built out westward into the former lake basin; these intrusions alternated with periods when the lake waters intruded eastward. The result is a complex interdigitation of lake sediments and stream sediments.

The major basis for correlating the various strata is the presence of distinctive strata consisting of tuffs; the volcanic material has periodically washed into the lakeshore basin from the terrain to the north and to the east. Some of the tuff beds are widespread, and some are not. The uncertainties of correlation between tuff layers in different locations are greatest in the Koobi Fora and Allia Bay areas; these are the areas farthest from the erosional sources of the volcanic ash. At the same time volcanic rocks can be dated by means of isotope measurements, which makes the ash strata particularly important.

Jack Miller and Frank J. Fitch have conducted most of the dating studies of the tuff (based on the decay of radioactive potassium into argon). Independent chronological data are also available from studies of the magnetic orientation of some particles in the volcanic ash and studies of fission tracks in bits of zircon in the ash. The various measurements are by no means unequivocal, but it can be stated as a generality that the location of a fossil find below one such layer of tuff and above another at least establishes a relative chronological position for the fossil even if its precise age remains in doubt.

There are five principal tuff marker layers. The earliest, which provides a boundary between the Kubi Algi sediments below it and the Koobi Fora sediments above it, is the Surgaei tuff. The next layer of tuff divides the lower member of the Koobi Fora sedimentary formation approximately in half; this is the Tulu Bor tuff, 3.2 million years old. The next layer, the KBS tuff, is named for the exposure where it was first recognized: the Kay Behrensmeyer site. It marks the boundary between the lower and upper members of the Koobi Fora Formation, and its exact age is debated.

Above the anomalous KBS tuff the next marker layer is the Okote tuff, which divides the upper member of the Koobi Fora sediments approximately in half. The Okote tuff is between 1.6 and 1.5 million years old. The fifth and uppermost marker layer, roughly indicating the boundary between the Koobi Fora sediments and the overlying Guomode sediments, is the Chari-Karari tuff; it is between 1.3 and 1.2 million years old.

Most of the early hominid fossils discovered in East Turkana are sandwiched between the Tulu Bor tuff at the most ancient end of the geological column and the Chari-Karari tuff at the most recent end. Twenty-six specimens, including a remarkable skull unearthed in 1972, designated KNM-ER 1470, come from sediments that lie below the KBS tuff. (The designation is an abbreviation of the formal accession number: Kenya National Museum-East Rudolf No. 1470.) Another 34 specimens, including several skulls, come from sediments that lie above the KBS tuff but below the Okote tuff. Unfortunately for those interested in measuring the rates of evolutionary change in hominid lineages, the difference between the oldest and the youngest proposed KBS-tuff dates is some 1.3 million years.

That span of time far exceeds the one allotted to the whole of human evolution not many years ago. Even by today's standards the KBS-tuff discrepancy is enough to allow uncomfortably different evolutionary rates for various hypothetical hominid lineages. So far, evidence of other kinds has not resolved the issue. For example, our colleagues John Harris and Timothy White, who have conducted a detailed study of the evolution of pigs throughout Africa, suggest that the more recent date for the KBS tuff would best suit their fossil data. At the same time the fission-track studies of zircons from the KBS tuff indicate that the older dates are correct. For the time being we must accept the fact that the KBS tuff is either about 2.5 million years old or somewhere between 1.8 and 1.6 million years old.

Relative and absolute chronology apart, other kinds of investigation are increasing our knowledge of the different environments that were inhabited by the East Turkana hominids. For example, most of the hominid specimens can in general be placed either in the genus *Australopithecus* or in the genus *Homo*. Behrensmeyer, Findlater and Bowen are engaged in microstratigraphic studies that have enabled Behrensmeyer to associate many of the specimens with a specific environment of sedimentary burial. Preliminary analyses indicate that the specimens identified as *Homo* were fossilized more commonly in lake-margin sediments than in stream sediments, whereas the specimens identified as *Australopithecus* are equally common in both sedimentary environments. Facts such as these promise to be of great help in reconstructing the lives of early hominids. In this instance the chance that an organism will be buried near where it spends most of its time is greater than the chance that it will be buried farther away. Thus Behrensmeyer has hypothesized that in this region of Africa early *Homo* exhibited a preference for living on the lakeshore.

Because of the unusual circumstances in this badlands region it will be useful to describe how the hominid fossils have been collected. The initial process is one of surface prospecting. The Kenyan prospecting team is led by Bwana Kimeu Kimeu; its job is to locate areas where natural erosion has left scatterings of mammalian bones and teeth exposed on the arid surface of the sedimentary beds. Kimeu is highly skilled at recognizing even fragmentary bits of hominid bone in the general bone litter present in such exposures.

Once the presence of a hominid fossil is established by the prospectors, one of the project geologists determines its position with respect to the local stratigraphic section and records the location. Thereafter one of two procedures is generally followed. If the bone fragment has been washed completely free from the sedimentary matrix that held it, the practice is to scrape down and sieve the entire surrounding surface area in the hope of recovering additional fragments. As the scraping is done a watchful eye is kept for fragments that might still be in situ, that is, partly or entirely embedded in the rock.

If the initial discovery is a fossil fragment in situ, the procedure is different. Excavation is begun on a near microscopic scale, the tools being dental picks and brushes. The Turkana hominid fossils are often so little mineralized that a preservative must be applied to the bone as excavation progresses in order to keep it from fragmenting further. Indeed, sometimes the preservative fluid must be applied with painstaking care because the impact of a falling drop can cause breakage.

After excavation each site is marked by a concrete post inscribed with an accession number provided by the Kenya National Museum. The next task, usually undertaken in the project laboratory in Nairobi, is piecing together the specimens. This is rather like doing a three-dimensional jigsaw puzzle with many of the pieces missing and no picture on the box. Any adhering matrix is now removed under the microscope, most often with an air-powered miniature jackhammer. (Cleaning with acid, a common laboratory method, is out of the question because the fossil bone is less resistant to the acid than the matrix.) Finally, the hardened pieces of bone are reconstructed to the extent possible by gluing adjacent fragments together.

Can the East Turkana collection be considered representative of the hominid populations that occupied the area more than a million years ago? The biases that affect fossil samples are many. For example, circumstances may result in the preservation of only some parts of certain individuals. Or a particular specimen may be severely deformed by pressure during its long burial. One bias that is easy to recognize in the East Turkana collection is a disproportionate number of lower jaws of the early hominid *Australopithecus robustus*. This hominid had powerful jaws and unusually large teeth; its lower jawbone is particularly massive. The relative abundance of these jaws and teeth in the East Turkana sediments probably results more from their mechanical strength, and thus their enhanced ability to survive fossilization, than from any preponderance of *A. robustus* individuals in the population.

Another example of bias in the hominid-fossil collection is the disproportionate representation of different parts of the skeleton. Teeth are by far the hardest parts, and so it is not surprising to find that teeth account for the largest fraction of the East Turkana sample. In contrast, vertebrae and hand and foot bones are rarely found. Can this bias be attributed to the destructive processes associated with burial and exposure alone? It seems only logical to take into account a third process: carnivore and scavenger feeding on the hominid bodies before the sediments covered them.

In developing such hypotheses we must keep in mind a number of fundamental questions. One question is: How many different species of early hominids were there in East Turkana? Another question is whether those species, whatever their number, existed over long periods of time or were replaced by other species. Again, do any or all of the species show signs of evolution during this interval of perhaps 1.5 million years or perhaps only 700,000 years? If there was any evolutionary change, what was its nature? How did the early hominids feed themselves? Were they relatively low-energy herbivores or relatively high-energy omnivores? If indeed several species were present at the same time, did each occupy a distinct ecological niche? What kept the niches separate?

The questions do not end here. Other questions, more specifically anatomical, also call for answers. Do the hominid fossils possess any morphological attributes that might be correlated with the archaeological record of tool use and scavenger-hunter behavior in the area? For example, can we detect any evidence of significant brain evolution during the period? Are there any morphological changes suggestive of altered patterns of locomotion or hand use that might shed light on the origins of certain unique human attributes? (For the purposes of this discussion we define these attributes as including not only walking upright and making use of tools but also an enlarged brain and the ability to communicate by speaking.)

These questions and many more can be answered only after the first question in the series is disposed of. Basically taxonomic in nature, it asks how many species are represented in the East Turkana fossil-hominid assemblage. We have already said that in general two genera were present: *Australopithecus* and *Homo*. How may the two be subdivided?

The answer to this basic and far from simple question is not an easy one. Conspiring against a clear-cut response are such factors as the smallness of the sample, the fragmentary condition of the individual specimens, the fact that even among individuals of the same species a large degree of morphological variability is far from uncommon and, under this same heading, the fact that a great deal of variation is often found between the two sexes of a single species. Also not to be neglected is the fallibility of the analyst, who is prone to human preconceptions. For example, the very order of discovery of the East Turkana hominids has affected our hypotheses, and we have had to chop and change in order to keep abreast of later discoveries.

It is illuminating in this connection to review the sequence of hominid discoveries in East Turkana. The first specimens to be identified were individuals of the species *Australopithecus robustus*. Fossil representatives of this species were first found at sites in South Africa decades ago. Characteristically they are large of face and massive of jaw; the molar and premolar teeth are very large, although the incisors and canines are small, about the same size as the front teeth of modern man. Although the facial skeleton is large, the brain case is relatively small: the average cranial capacity is about 500 cubic centimeters, compared with the modern human average of 1,360 cc. Because the chewing muscles were evidently of a size commensurate with the large cheek teeth and massive jaws, many *A. robustus* individuals have not only extremely wide-flaring cheekbones but also a bony crest that runs fore and aft along the top of the brain case to provide a greater area for the attachment of chewing muscle.

Specimens of *A. robustus* have also been found in East Africa, most notably by Louis and Mary Leakey at Olduvai Gorge. The East African examples are on the whole even larger than those from South Africa, and their cheek teeth are more massive. A well-known example is "Zinjanthropus," an Olduvai cranium now accepted by most scholars as being closely related to the *A. robustus* specimens from South Africa. (Some scholars, it

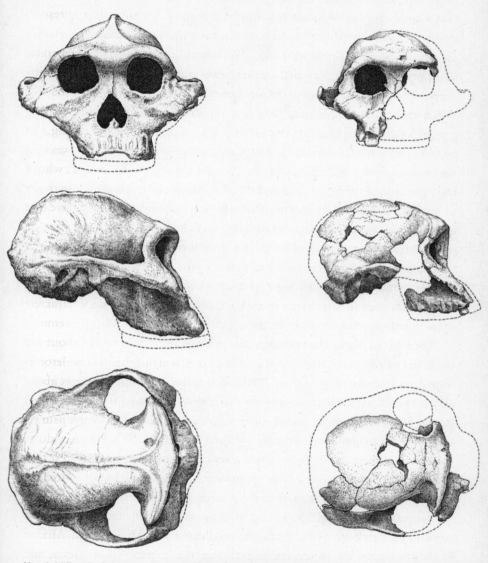

Hominid Fossils found early in the process of collecting in East Turkana are illustrated. At the left is Kenya National Museum-East Turkana Fossil (KNM-ER) accession no. 406, a robust cranium with well-preserved facial bones. At right is KNM-ER 732, a fragmented cranium that has little of the face preserved and is less robust than KNM-ER 406. Both specimens are placed in the genus *Australopithecus*.

should be noted, still assign "Zinjanthropus" to a related species of *Australopithecus, A. boisei*.) the East Turkana deposits have been found to contain a good number of fossils that can be placed in this hominid species.

In the early investigations at East Turkana the skulls of certain smaller and less robust hominids were also discovered. Indeed, one such cranium, deformed by crushing, turned up near (although not in the same stratigraphic horizon as) a robust *Australopithecus* cranium: KNM-ER 406. When this crushed specimen was first discovered, it could not easily be given a taxonomic position. The finding of a second gracile (as opposed to robust) specimen, however, suggested to us that male-female dimorphism might account for both kinds of cranium. In the second specimen most of the right side of the brain case and facial skeleton and part of an upper-jaw premolar tooth and the roots of the molars were preserved. It is evident that although this individual is substantially less robust than KNM-ER 406, its premolar and molar teeth were only a little less massive than those of the robust one. If among the species *A. robustus* the morphological differences between males and females were as great as they are among gorillas, then the robust, crested specimens from East Turkana could be males and the more gracile specimens could be females.

The age of these *Australopithecus* specimens is substantially greater than that of any previously uncovered in East Africa but their discovery presented no taxonomic problems. This happy state of simplicity came to an end in 1972 with the discovery of the cranium KNM-ER 1470. Bwana Bernard Ngeneo came across it on an exposure of older sediments belonging to the lower member of the Koobi Fora Formation. When he found the specimen, all that could be seen was a scattering of bone fragments on the rock surface. The fragments were relatively fragile, which led us to assume that they had been washed out of the matrix quite recently.

The specimen KNM-ER 1470 is a large, lightly built brain case with a considerable amount of the facial skeleton preserved. Our colleague Ralph L. Holloway. Jr., has determined that its cranial capacity is about 775 cc. The facial skeleton is very large, and the proportions of the front and cheek teeth are indicated by the preserved tooth sockets and by both the sockets and the broken roots of the molars. The proportions are the reverse of those for *A. robustus;* the incisors and canines are very large and the premolars and molars are only moderately large. Even though the tooth size suggests a formidable chewing apparatus, the brain case shows no sign of a crest for the attachment of heavy chewing muscles. Similarly, the cheekbones, although incomplete, do not suggest the same great width of face that is characteristic of *A. robustus.*

Much has been written about the significance of KNM-ER 1470. We believe that certain hominid specimens found at Olduvai Gorge in broken and fragmentary condition are examples of the same kind of skull. If it is necessary to decide on a taxonomic term for these hominids, the species name may well turn out to be *habilis. (Homo habilis* is the name that was given to an early species of the genus *Homo* by Louis Leakey and his colleagues John Napier and Phillip V. Tobias.

It was at about the time of the discovery of KNM-ER 1470 that we and our colleagues began to disagree as to the taxonomic position of certain well-preserved lower jaws from the East Turkana region. The initial source of disagreement was a small cranium: KNM-ER 1813. It is the cranium of a small-brained hominid with the average *Australopithecus* cranial capacity: 500 cc. It has a relatively large facial skeleton and palate. The upper teeth preserved with the palate are comparatively small, however, and bear a striking resemblance to the teeth of one of the *Homo habilis* specimens from Olduvai, OH-13. It happens that in the initial controversy over the original *H. habilis* specimens, OH-13 represented a species that even skeptics agreed was nearly, if not actually, identical with the species *Homo erectus,* a member in good standing of the genus *Homo* that was first recognized in Java and northern China.

The resemblances between KNM-ER 1813 and OH-13 go further than their teeth. In all the parts that can be compared—the palate as well as the teeth, much of the base of the skull and most of the back of the skull—the two specimens are virtually identical. This leads us to believe the usual reconstructions of OH-13, which have assumed that the specimen had a large cranial capacity and an *erectus*-like skull, are in error. That is not all. The mandible of OH-13 was preserved; its small size and the details of the teeth were major components of the evidence leading to the conclusion that *habilis* was near the *erectus* line. The comparable lower jaws we have found at East Turkana, we can now see, make it clear that the OH-13 mandible could just as well have been hinged to a small-brained, thin-vaulted skull like that of KNM-ER 1813.

In the 1975 season we discovered a remarkably complete cranium: KNM-ER 3733. The find showed unequivocally that a member of our own genus was present in East Turkana when the early strata of the Koobi Fora upper member were formed. The skull bears a striking resemblance to some of the *Homo erectus* skulls found in the 1930's near Peking and is certainly a member of that species. The brain case is large, low and thick-boned. Its principal part is formed by the projecting occipital bone, and its cranial capacity is about 850 cc. The brow ridges jut out over the eye orbits, and a distinct groove is visible behind them where the frontal bone rises toward the top of the vault.

KNM-ER 3733 has a small face tucked in under the brow ridges. The sockets of some of the upper teeth are preserved. The missing front teeth were relatively large, but the back teeth (some of them still in place) are of only modest proportions. The third molars are among the missing teeth, but the evidence is that they were quite reduced in size. They had been erupted long enough for them to wear grooves in the second molars in front of them, yet the bone forming the sockets indicates that their roots were very small. The point is important because a diminution in the size of this tooth is a common phenomenon in modern human populations.

This fine example of *Homo erectus* from East Turkana predates the example of the same species found at Olduvai Gorge by half a million years and is about a million years older than the examples from northern China. And KNM-ER 3733 is not alone. Another *erectus* specimen was found shortly afterward in Area No. 3 of the Ileret fossil beds. This is KNM-ER 3883. The Ileret specimen is from approximately the same geological horizon as KNM-ER 3733. It has much the same cranial conformation, but its brow ridges, facial skeleton and mastoid processes are somewhat more massive. The cranial capacity of KNM-ER 3883 has not yet been determined, but there is no reason to expect that it will be much different from that of 3733. The similarity of the two East Turkana specimens to specimens from far away that are very much younger strongly suggests that *Homo erectus* was a morphologically stable species of man over a span of at least a million years.

Leaving aside the problems presented by various fragmentary specimens, how can the new hominid finds from East Turkana be assessed taxonomically? One might simply suggest a series of normal taxonomic assignments in the light of what we see at present, acknowledging that as in the past the assignments are likely to be changed. We shall not do so here because we now view the problem of taxonomic assignment in a slightly different way.

Considering only the fossils from the upper member of the Koobi Fora Formation, we think that three species are present. Sorted out in this way, none of the specimens within each group shows more variability in brain size and chewing apparatus than we see among living anthropoids. Although the fossil record from the lower member of the Koobi Fora Formation is far less rich than that from the upper member, a similar three-species hypothesis could also be advanced with respect to the specimens found in it. In this hypothesis the third species in addition to the robust and gracile ones would be represented by the *habilis* specimens.

Several consequences follow from our probability assessments. For example, in our view the demonstration at East Turkana that *Homo erectus* was contemporaneous with some of the largest representatives of *Australopithecus robustus* amounts to a disproof of the single-species hypothesis. We believe both *H. erectus* and *A. robustus* had essentially human characteristics. If this is the case, it follows that they occupied separate ecological niches. It would seem either that one of the species did not possess culture and yet still developed human characteristics or that the argument that the advantage of culture would give the cultured hominid dominance within a very wide ecological niche is flawed.

We prefer the first alternative, and we would nominate *Australopithecus* for the role of the hominid without culture. Accepting this alternative requires that we keep searching for the natural-selection pressures that have been responsible for producing the basic human attributes.

Where did *Homo erectus* come from? Some have suggested that the species arose in Asia and migrated to Africa. This seems to us an unnecessarily complicated hypothesis. For one thing, it neglects *habilis*. Worse, it implies that a population of these large-brained hominids, who presumably made the stone tools found in the early East Turkana strata, evolved independently in Africa at the right time to fit into an ancestor-descendant relation with *Homo erectus* and then came to an abrupt halt, without playing any further part in human evolution.

It is our view that the *habilis* populations are directly antecedent to *Homo erectus*. If the earlier range of dates for the strata where the *habilis* specimen KNM-ER 1470 was found proves to be correct, then the transition from *habilis* to *erectus* could have been a gradual one, spanning a period of well over a million years. If the later dates are correct, then the transition must have been very rapid indeed.

These studies and others aim at reconstructing as much as can be reconstructed about the biology of these very early hominids in the hope of determining just what it was that made us human.

The Omnivorous Chimpanzee

Geza Teleki

It is widely believed that apes and monkeys are vegetarians, and that man is alone among the primates in preying on other animals. The assumption has influenced a number of hypotheses about human evolution that were framed in the days when scarcely any of man's primate relatives had been studied in the wild. For example, it has been suggested that the pursuit of game and the consequent social sharing both of the hunt and of the kill were key factors in the divergence of the earliest hominids from the rest of the primate line.

Today, after years of field observations of ape and monkey behavior, it is quite clear that man is not the only primate that hunts and eats meat. Many other primates are omnivorous. One in particular—the chimpanzee—not only cooperates in the work of the chase but also engages in a remarkably socialized distribution of the prey after the kill. The chimpanzees whose predatory behavior has been most closely observed are semi-isolated residents of the Gombe National Park in western Tanzania. The area, formerly known as the Gombe Stream Chimpanzee Reserve, is where Jane van Lawick-Goodall began her notable long-term field study of chimpanzees in 1960. I myself spent 12 months watching the predatory behavior of these apes in 1968–1969.

Gombe Park covers some 30 square miles and has an estimated population of 150 chimpanzees. All belong to the eastern chimpanzee subspecies *Pan troglodytes schweinfurthii*. Goodall and her colleagues quickly came to know on sight some 50 individual apes that lived in a 10-square-mile zone centered on Kakombe Valley. This zone became the main study area, and its chimpanzees were made the subjects of daily records.

It was soon apparent to the observers that the chimpanzee population was organized in a social hierarchy headed by a single adult male that was senior to all the other males, adult and subadult. Most adult males and even some of the subadults, in turn, usually outranked the female chimpanzees, regardless of the females' age. The females' behavior in many situations showed, however, that there was an independent hierarchy among them as well. Between 1960 and 1970 the position of senior male, or "alpha," was occupied successively by two different adults. During my 12 months of observation the position was held by a chimpanzee Goodall had named Mike.

Sherwood L. Washburn and Irven DeVore of the University of California at Berkeley were among the first to observe primates eating meat. Studying olive baboons and yellow baboons in Kenya in 1959, they saw these ground-dwelling primates kill and eat newborn antelopes. Not long afterward Goodall found that the Gombe chimpanzees were omnivorous. Their diet included insects, lizards, birds' eggs and fledgling birds, young bushbucks and bushpigs, blue monkeys, redtail monkeys, colobus monkeys and infant and juvenile baboons. Meanwhile observations of primates in the wild around the world revealed that an omnivorous diet was far from unusual. By the end of the 1960s the list of primates that feed on animals larger than insects had grown to include two chimpanzee populations outside Gombe Park, two species of baboons in addition to the olive baboon and the yellow baboon, vervet monkeys in Africa, macaques in Japan and even woolly monkeys and capuchin monkeys in the New World. Only baboons and chimpanzees, however, are known to actively seek out and pursue their prey.

In the decade between 1960 and 1970 Goodall and others, myself included, were able to note that the 50-odd chimpanzees in the vicinity of the Kakombe Valley field station killed and ate no fewer than 95 individual mammals and attempted to capture another 37 that escaped. In 46 of the predatory incidents the kills were actually witnessed. Another 38 kills were known through examination of the chimpanzees' feces, and recognizable fragments of 11 additional prey animals were carried by the chimpanzees to within sight of an observer.

During my year at Gombe Park I witnessed 30 episodes of predation, 12 of them successful. Kills during the decade, including those I observed, averaged a little more than nine mammals per year. The fact that I witnessed 30 episodes and 12 kills in a 12-month period is less likely to mean that the period was an above-average one for predation than that in the other years some kills and many episodes probably went unnoticed. In terms of predation on baboons, however, the period of my observations was clearly not average. Goodall's records for the decade include identifications of 56 of the 95 prey animals. Primates were in the majority: 14 colobus monkeys, 21 baboons, one blue monkey and one redtail monkey. Of the 19 other mammals identified 10 were young bushpigs and the rest were young bushbucks. In contrast, of the kills I observed in 1968–1969 the prey animal on 10 out of 12 occasions was a baboon. Moreover, all 18 unsuccessful episodes I witnessed involved baboons as prey animals.

The prey species I have mentioned are very nearly the only mammals available to the Gombe chimpanzees, so that what might appear to be preference in reality demonstrates the diversity of the apes' predatory efforts. Indeed, the only limit to chimpanzee predation that I observed was the size of the prey animal. There is no evidence that chimpanzees capture or even pursue an animal that weighs more than about 20 pounds. For example, most captured baboons were infants or juveniles with an estimated weight of 10 pounds or less. Similarly, the bushbucks and bushpigs that the chimpanzees kill are either newborn or very young. Few of the adult mammals killed by the Gombe chimpanzees (colobus monkeys, blue monkeys and redtail monkeys) weighed as much as 20 pounds.

The prevalence of baboons as prey in Gombe Park must be due to the fact that baboons are the primates the chimpanzees most frequently encounter. Both species live mainly on the ground, both travel the same trails through grassland and forest, and both eat many of the same foods and visit the same foraging areas. Each species may displace the other from special feeding sites, such as trees bearing fruit, and may even interact competitively over access to favored foods. At the same time baboons and chimpanzees engage in amicable interactions such as grooming and play. During my 12 months at Gombe I saw play groups made up of young chimpanzees and baboons at the field station almost daily. On two occasions I saw adult male chimpanzees suddenly pursue and capture a young baboon that until shortly before had participated in a mixed play group that was being tolerantly observed by female chimpanzees and baboons sitting a short distance away.

The coexistence of both social and predatory interactions between chimpanzees and baboons at Gombe may be anomalous. Elsewhere in Africa adult male baboons are reported to defend their troop against threatening carnivores, even to the extent of seriously injuring such formidable adversaries as large leopards. Indeed, I observed the baboons of one Gombe troop kill a 35-pound African wildcat that appeared to have threatened them.

The Gombe baboons' response to predatory chimpanzees appears to be far more ambivalent. I have seen an adult male baboon sit unchallenged shoulder to shoulder with three adult male chimpanzees that were dividing the carcass of a bushbuck, while other baboons and chimpanzees searched together through the underbrush for fallen pieces of meat. At the opposite extreme I have seen a single chimpanzee stand upright and run into the midst of a baboon troop, where it selected, pursued and captured a young baboon while exhibiting complete indifference to the numerous adult male baboons that "mobbed" it, threatening it, slapping it and even leaping on its back. Nor was this an unusual event. The Gombe records show that predatory chimpanzees are rarely more than scratched by the baboons that undertake the defense of a troop member. A severe injury to a chimpanzee predator has never been reported.

Perhaps it is both the frequency and the variety of social interactions between the Gombe chimpanzees and baboons that are responsible for the ambivalence of the baboons' response. The use of the same trails and foraging areas, the mutual recognition of many communicative signals and the grooming and play among the young of both species may have resulted in the development of an interspecific "communal" atmosphere, so that the baboons respond to the chimpanzees much as they might respond to other baboons. That something of this kind is possible, at least in reverse, is apparent from the following observation. An adult baboon was fatally injured by another member of the troop; the fight was witnessed by several chimpanzees. Instead of considering the dead baboon fair prey, some of the chimpanzees inspected the carcass, touched it and finally groomed it. The same kind of curiosity has been observed when chimpanzees are confronted with a dead chimpanzee.

Insofar as it has been observed among the Gombe chimpanzees, hunting behavior is an exclusively adult activity and is almost exclusively a male one. Adult females have occasionally been observed pursuing and even capturing prey, but in every instance no adult males were in the vicinity at the time. The males may hunt alone, or two or more males may coordinate their actions. I once witnessed five males working together to surround three baboons that had taken shelter in trees; the movements of the chimpanzees were plainly cooperative.

A predatory episode that results in a kill consists of a sequence of three events, each marked by its own characteristic activities. The first of the three I shall call "pursuit," even though the distance covered is sometimes only a few feet and the time elapsed is seconds. The second event, "capture," is a brief period that ends with the initial dismemberment of the prey. The third and longest event, "consumption," involves highly structured activities. On one occasion I observed a consumption period that lasted nine hours and involved 15 chimpanzees.

The mean duration of the 12 successful predatory episodes I witnessed was a little less than four hours; the shortest episode lasted an hour and 45 minutes. An unsuccessful episode is of course much briefer. The mean duration of the 18 I witnessed was 12 minutes. The Gombe chimpanzees spend more than 90 percent of the total time they devote to predation in sharing and eating the prey.

I did not always have the prey animal under observation at the start of a predatory episode, so it was not always clear to me what had initiated the pursuit. When both the prey and the predator were in view, it was apparent that the chimpanzees perceived and often selected their prey before starting the pursuit. The indicative changes in the hunter's posture and expression were so subtle, however, that it is difficult to specify them.

Many episodes began when the male chimpanzees were relaxed, for example dozing or grooming or resting after they had eaten large quantities of fruit. When the chimpanzees were either interacting intensively among themselves or interacting aggressively with some other species of primate, predatory episodes were uncommon. When a baboon was the prey animal, one apparent stimulus to predation was the vocalization of infant or juvenile baboons. Associated with both aggression and distress, these sounds frequently occurred during the young baboons' play sessions or when a young baboon sought to return to its mother. In more than half of the 28 episodes involving baboons that I witnessed, the crying of young baboons preceded any evidence of predatory interest on the part of the chimpanzees.

As I have indicated, one form of pursuit is simple seizure, an explosive act that lacks obvious preliminaries. The chimpanzee takes advantage of a fortuitous situation to make a sudden lunge and grab the prey animal. In this respect seizure is a kind of instantaneous capture that resembles "opportunistic" feeding on immobile prey (for example a fledgling bird or a "frozen" newborn antelope), the kind of meat-eating frequently noted among lower primates. I was able to witness this form of pursuit in 22 of the 30 predatory episodes I observed; seven involved simple seizure, and three of the lunges were successful.

Two other forms of pursuit are practiced by the Gombe chimpanzees. One is chasing, which may involve a dash of 100 yards or more. The other is stalking, a cautious and painstaking process that can last more than an hour. Both have the appearance of being more premeditated and controlled than simple seizure. Moreover, on occasion both clearly involve a strategy and maneuvers aimed at isolating or cornering the selected prey. Eleven of the episodes that I observed involved chasing and four involved stalking. Six of the chases were successful, but all the stalking efforts failed.

At times, particularly in the early morning and in the evening, the Gombe chimpanzees are quite vociferous. That is not the case when they are in pursuit of prey. Regardless of the time of day or the number of chimpanzees involved in the chase, all remain silent until the prey is captured or the attempt is broken off. This means, of course, that the hunters do not coordinate their efforts by means of vocal signals. Neither did I observe any obvious signaling gestures, although cooperation in movement and positioning was evident. It is noteworthy that the leading position during the chase frequently shifted from one hunter to another regardless of the chimpanzees' relative social rank. Evidently individual chimpanzees do not compete with one another for the most advantageous position during the pursuit of prey.

After the prey animal has been maneuvered within reach begins the central event in the predatory sequence: capture. It usually lasts less than five minutes and exhibits three consecutive stages: acquisition, killing and initial division. The first stage is very much like simple seizure; it consists of a final lunge and grab when the distance between the predator and the prey has been reduced to a yard or less. If the chase has been a cooperative venture, more than one chimpanzee may catch the prey. The instant of acquisition is usually signaled by a sudden outburst of vocalization; not only do the cries end the silence of the hunt but their volume and pitch serve to draw other chimpanzees from distances of a mile or more.

The killing stage is normally brief. If the prey is in the grasp of a single chimpanzee, the chimpanzee may bite the back of the prey's neck or twist its neck in both hands. Alternatively the chimpanzee may stand upright, grasping the prey by its legs, and strike its head and body against the ground or a tree trunk. If the prey is caught by more than one chimpanzee, it may be torn apart as each captor tugs on a different limb; in this way killing and dividing are accomplished simultaneously.

The final stage of capture, initial division of the carcass, is an activity that is unrelated to the "meat-sharing" that comes later. For at least a brief period after the killing the carcass appears to be "common property." If chimpanzees other than the captor or captors have arrived within reach of the prey, they are free to try to grab a part of the carcass without risk of retaliation from the similarly occupied captors. For example, I have seen six chimpanzees, four of them postkill arrivals at the scene, divide among them the arms, legs and trunk of a young baboon. Even under these conditions aggressive interactions are rare during the time of initial division. The few incidents I witnessed tended to be mild, and even male chimpanzees of high social status showed nearly complete tolerance for others.

If the captor is able to hoard the prey for several minutes by moving away from the kill site, the common-property character of the prey animal lapses. Chimpanzees that have been attracted to the site are no longer likely to try to tear off a part of the carcass. In the event of a capture by more than one chimpanzee, the same "hoarders' rights" apply equally to the major portions of the carcass. The chimpanzees that share in the initial division then move off, usually no more than a dozen yards apart, each hoarding its piece of the carcass; those that did not participate in the initial division begin to congregate in "sharing clusters" around each possessor of a major fragment. Formation of the sharing clusters initiates the third, and socially the most significant, event in the predation sequence: the consumption period.

Considering the length of time devoted to consumption, the small size of the prey animals and the number of chimpanzees that congregate in sharing clusters, the conclusion is almost inescapable that social considerations and not merely nutritional ones underlie the Gombe apes' predatory behavior. For one thing, many predatory episodes are initiated soon after the chimpanzees have consumed large quantities of vegetable foods. For another, no chimpanzee at Gombe has ever been observed to capture and privately consume a mammal, however small, if other adult chimpanzees were present to form a sharing cluster. In the decade covered by the Gombe records exactly two unshared kills were observed. The events were simultaneous: two adult female chimpanzees happened to encounter two small bushpigs and each ate one of them. No other adult chimpanzees were present at the dual kill.

Before a sharing cluster disperses, all the prey animal's carcass will have been consumed: skin and hair, bones, bone marrow, eyeballs and even teeth. The brain is evidently the preferred portion. Although I observed frequent sharing of other parts of the carcass, not once in 12 months did I see one ape yield the brain or any part of it to another. A chimpanzee sometimes removes the brain tissue by pushing a finger into the natural opening where the skull joins the backbone. More often it opens a hole in the prey's forehead with its fingers and teeth and then scoops and sucks the cranial cavity clean.

The prey animal's brain may be eaten first or last. Otherwise when a single chimpanzee captor deals with its prey, the dismantling sequence usually begins with the removal and consumption of the viscera. Next the rib cage is cleaned and sectioned, the chimpanzee using its teeth, hands and feet to tear the skin and break the bones. The prey's limbs are consumed last. The process is thorough; a careful inspection of a kill site following the chimpanzees' departure yields nothing but a few tiny fragments of the prey.

I have seen chimpanzees in a sharing cluster make use of three methods of getting meat. One that involves a minimum of interaction with the possessor of the meat is to retrieve dropped or discarded bits of the prey that fall to the ground. Such retrieval is usually the activity of subadult chimpanzees and adult females that evidently prefer not to approach the possessor directly.

The second method is simple taking: tearing off a portion of the prey or even seizing bits of meat from the possessor's hands or mouth. This direct approach is often used when the possessor is a female and the taker is one of its offspring. Male possessors are most likely to tolerate meat-taking by another adult male, particularly a sibling, by a female who is sexually receptive or by a female of high social status. The direct approach often finds taker and possessor calmly chewing on the same piece of meat, and I have seen as many as three sharing-cluster apes so engaged without discouragement from the possessor. Subadults rarely attempt the direct approach.

The third course of action, which involves specific behavioral patterns, I call requesting. Meat can be requested in the following ways. The requester can approach the possessor closely, face to face, and peer intently at the possessor or at the meat. Alternatively the requester can extend a hand and touch the possessor's chin or lips or touch the meat. The requester can also extend a hand, open and palm up, holding it under the possessor's chin. The requester may accompany each of the gestures with soft "whimper" or "hoo" sounds. Chimpanzees of virtually any age and of either sex request meat with this repertory of gestures. The youngest I observed doing so was an 18-month-old infant; the most important socially was "alpha" Mike, the top-ranking chimpanzee at Gombe Park.

If the possessor's response to a request is negative, the denial is indicated by ignoring the requester or turning away, by pulling the meat out of reach, by moving to a less accessible place, by vocalizing, by gesturing and occasionally by pushing the requester away. The possessor responds in the affirmative by allowing the requester to chew on the meat or tear off a portion, or by dropping a piece of meat into the requester's upturned hand. Of the 395 requests I observed, 114 were rewarded. Occasionally the possessor will detach a considerable portion of meat and with outstretched hand offer it to the requester. I saw this done only four times during my year at Gombe. On one of those occasions the possessor was holding an entire carcass. An adult male in the sharing cluster had been requesting meat persistently, and the possessor at last divided the carcass in two and handed half to the requester.

When chimpanzees are sharing meat, their behavior is generally relaxed and uncompetitive. The pieces of the prey animal are consumed in a leisurely fashion and are evidently relished. I saw very few hostile interactions between apes with meat and apes with none. Individuals in a sharing cluster that had waited in vain for a long time would sometimes threaten or chase other chimpanzees that were also waiting, but they never made such a move against the chimpanzee with the meat.

High social rank is apparently no guarantee of success in requesting meat. Even though the possessor may be a relatively low-ranking adult, a high-ranking adult in the sharing cluster will approach the possessor with the same repertory of gestures that the other requesters use and will make no effort to assert superior social status aggressively. I once observed "alpha" Mike request meat for several hours from a subordinate male; Mike received nothing for his pains. Twice during my year of observing I saw meat seized from a possessor by a surprise dash-and-grab maneuver that may have had hostile overtones; the possessor's only retaliation was to scream and wave its arms. In the total of 43 hours of meat-sharing that I observed not once did two chimpanzees fight over possession of meat.

To what extent do the Gombe findings and similar observations elsewhere suggest a modification of present views concerning human evolution? The theme of hunting as a way of life has been central to many of the hypotheses that attempt to trace the evolution of human behavior and human social organization.

It seems possible to me that predation developed among primates long before the advent of ape-men. To an omnivorous primate that enjoys complete mobility in a three-dimensional habitat other primates are perhaps the most readily accessible prey.

Most of the mammals the Gombe chimpanzees kill and eat are primates. The fossil remains of *Australopithecus* in South Africa have been found in association with the damaged skulls of baboons. The association at least suggests the possibility that *Australopithecus* occasionally preyed on the primates that shared its range.

Suppose that predation, cooperative hunting and socially structured food-sharing did become habitual among certain primates long before the first hominids arose. If that were the case, it would throw doubt on a number of current evolutionary hypotheses. For example, the sequence of erect posture, free hands and tool use as prerequisites to the emergence of hunting behavior would no longer appear to be valid. Moreover, whether or not predation is a far more ancient primate behavior pattern than has been supposed, at least one hypothesis must be abandoned altogether. As the actions of the Gombe chimpanzees demonstrate, socially organized hunting by primates is by no means confined to man.

The primary significance of the Gombe chimpanzees' predatory behavior is not, however, dietary. What is far more important is the behavior that accompanies the predatory episodes: cooperation in the chase and the sharing of the prey. The more we learn about primate behavior, the smaller the differences between human and nonhuman primates appear to be. The observations at Gombe do much to reinforce this conclusion.

Chapter 2

Beasts of the Land

On the Origins of Subspecies

During the 1800s, whalers and seal hunters slaughtered the Galápagos giant tortoises for an easy supply of food. Those invaders and colonists also brought with them goats, rats and other animals that have eaten the tortoises' food, trampled their nests and attacked their hatchlings. The result, according to some researchers, is that three subspecies of the venerable reptile are now extinct, and a fourth has dwindled to a single known survivor. To stem this trend, conservationists have raised hundreds of hatchlings until the young tortoises were large enough to be introduced to the wild with minimal danger from predators. In a similar vein, authorities want to set free dozens of adult tortoises of unknown origins—many of them confiscated from poachers—but scientists have been unsure on which island to place each animal. DNA analysis could provide the answers.

The tortoises are currently being held at the Charles Darwin Research Station on Santa Cruz Island in the Galápagos archipelago. Conservationists there have been reluctant to release the animals back into the wild without knowing for sure where they came from because the islands have, according to some experts, evolved genetically distinct subspecies. The problem is that many of those subspecies are difficult to distinguish visually, and mixing them could lead to "unnatural" hybrids. Consequently, warns James P. Gibbs, a conservation biologist, "You really don't want to be tossing tortoises just anywhere."

So Gibbs, along with a team headed by geneticist Jeffrey R. Powell, has taken blood samples from hundreds of wild tortoises on the different islands. By analyzing the DNA of the samples, the researchers report they so far have found unique markers for all but four of the reputed 11 extant subspecies. (The taxonomy of Galápagos tortoises has been controversial; one debate questions the validity of at least several of the subspecies.) Using these signature sequences, the scientists plan to identify the home islands of many of the tortoises held in captivity.

A potential problem is that the captivity of the tortoises, especially with a mixture of subspecies, may have altered the natural behavior and instincts of the animals, which could prove disruptive if they are returned to the wild. One possible result is that a tortoise might no longer be able to follow an important migratory path on its home island. Such considerations, conservationists warn, make repatriation a tricky issue. "We can provide information on a particular tortoise's origin," Powell notes, "but there are a lot of questions about what to do with that information."

—*Alden M. Hayashi, editor,* Scientific American

Killer Kangaroos and Other Murderous Marsupials

Stephen Wroe

Dawn mist blankets the rain forest of Riversleigh in northeastern Australia, 15 million years ago. A bandicoot family emerges to dip snouts warily into a shallow freshwater pool. Their ears swivel, ever alert to a sudden crack or rustle in the undergrowth: drinking is always a dangerous activity. Suddenly, a dark, muscular form explodes from behind a nearby bush, colliding with a young bandicoot in one bound. The shaggy phantom impales its victim on long, daggerlike teeth, carrying the carcass to a quiet nook to be dismembered and eaten at leisure.

In nature, many animals will meet a violent death. So the sad end of one small bandicoot seems hardly worth mention. The demise of this little fellow would, however, have surprised most modern onlookers. Its killer was a kangaroo—the Powerful-Toothed Giant Rat-kangaroo (*Ekaltadeta ima*), to be exact.

In 20th-century Australia, warm-blooded predators are few and far between. Among our natives, the largest carnivores are the Spotted-Tailed Quoll (*Dasyurus maculatus*) and the Tasmanian Devil (*Sarcophilus harrisii*). (The doglike dingo, which also eats flesh, did not originate in Australia but was introduced by humans between 5,000 and 4,000 years ago.) The Spotted-Tail Quoll is a marsupial that weighs up to seven kilograms (15 pounds); it is also known as a native "cat" because of a passing resemblance to ordinary, placental cats. The Tasmanian Devil, another marsupial, is only slightly larger and looks like a lapdog with a fierce hyena's head. It is arguably the least fussy eater in the world and will devour an entire carcass, including the teeth. This odd pair is placed in the family Dasyuridae, which includes other native cats as well as far smaller, mostly insectivorous creatures called marsupial mice.

Some scientists have suggested that Australia has never supported a healthy contingent of large warm-blooded carnivores. Their evolution was

Largest marsupial lion (thylacoleo carnifex), 130 to 260 kilograms.

Largest marsupial wolf (thylacinus potens), maximum 45 kilograms.

constrained by poor soils and erratic climate for the past 20 million years or so. These constraints limited plant biomass, in turn restricting the size and abundance of potential prey animals. Instead, reptiles such as the seven-meter-long (23-foot-long) lizard *Megalania prisca*, which lived in Pleistocene times, took up the role of large terrestrial carnivores. Cold-blooded predators require less food than warm-blooded ones and so—the argument goes—were more likely to survive difficult conditions.

This claim is challenged by recent developments, notably spectacular fossil finds in Riversleigh, Queensland. A European naturalist, W. E. Cameron, first noted the presence of fossils at this remote site in 1900. But Cameron believed that the material he had seen was fairly young, less than two million years old. Moreover, Riversleigh's extreme inaccessibility—summer heat and monsoon rains allow excavations only in winter—persuaded paleontologists to neglect the locality for decades. In 1963, however, Richard Tedford of the American Museum of Natural History and Alan R. Lloyd of the Australian Bureau of Mineral Resources took a gamble and visited the site. They found the fossils intriguing and older than previously believed but fragmentary and hard to retrieve.

Still, their findings stimulated other expeditions to Riversleigh, and in 1983 Michael Archer, of the Australian Museum in Sydney, struck paleo pay dirt. In an idle moment at the site he looked down at his feet and saw a very large lump of rock that just happened to contain as many new species of Australian Tertiary mammals as had been described in previous centuries. Since then, new specimens, including large carnivores, haave emerged at a prodigious rate. Many are exquisitely well preserved, so much so that some could be mistaken for the remains of animals that died only weeks ago.

The ancient creatures appear to have been mostly trapped in limestone caves. Their bones, which were quickly and perfectly preserved by water rich in calcium carbonate, testify to a lost menagerie of beasts that were every bit as deadly as, but far stranger than, anything known today. Since 1985 nine new species from Riversleigh, each the size of the Spotted-Tailed Quoll or bigger, have more than doubled the tally of large Australian carnivores at least five million years old. This bestiary now includes two kinds of giant rat-kangaroo, nine species of marsupial "wolf," five species of marsupial "lion" and one native cat.

The giant rat-kangaroos (propleopines) are closely related to the Musky Rat-kangaroo. This tiny animal, still found in the rain forests of Queensland, weighs less than a kilogram—small enough to look like a rat. It eats a wide variety of plant stuffs and small animals, and alone among living kangaroos it cannot hop. A living fossil, it is the last and tiniest survivor of a family that included some fearsome, muscle-bound cousins. The giant rat-kangaroos ranged from around 15 to 60 kilograms in weight. Like their diminutive descendant, they probably walked on all fours.

Tasmanian Devil

The marsupial wolves (thylacinids) and marsupial lions (thylacoleonids) are so named because of their superficial physical resemblances to canines and felines, although they were more closely related to kangaroos. The last of the marsupial wolves, perhaps confusingly called the Tasmanian Tiger because of the stripes on its rump, was exterminated because of a largely undeserved reputation for preying on sheep. Like cats, the marsupial lions had short, broad, powerful skulls, and they probably filled similar ecological niches as well; their size ranged from that of a house cat to that of a lion. Although no fossils contain actual traces of a pouch, specialized features of the bones shared with living animals leave no doubt that all these creatures were marsupials.

Fearsome Forest

For much of the Miocene epoch (25 to five million years ago), Australia was carpeted in wall-to-wall green, and rain forest covered many areas that are now savanna or desert. These jungles were evolutionary powerhouses, nurturing a far greater diversity of life than any modern Australian habitat does. A day trip through one of these forests would have been filled with surprises, many of them potentially dangerous.

One would have been the Powerful-Toothed Giant Rat-kangaroo, among the most ancient of rat-kangaroos (another five species have been described from younger deposits). *E. ima* was also the smallest, weighing only about 10 to 20 kilograms. It is well represented by two nearly complete skulls. These fossils give us our best shot yet at understanding the feeding habits of the giant rat-kangaroos.

Because these animals descended from plant-eating marsupials, some controversy surrounds the interpretation of their biology. Nevertheless, authors agree that these distinctly uncuddly kangaroos included meat in their diets. Evidence supporting this hypothesis comes from both their skulls and their teeth.

In popular imagination, ferocious meat-eaters usually come with large canines. In the main this holds true, but there are some exceptions. Many humans consume a good deal of flesh—more than some so-called carnivores—but we have small canines, whereas in gorillas, which are vegetarians, these teeth are large. The real hallmark of a terrestrial mammalian killer is a set of distinctive cheek teeth used for cutting and shearing.

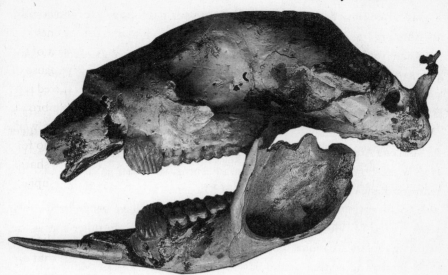

A fossil skull of the Powerful-Toothed Giant Rat-kangaroo displays the fearsome incisors and serrated carnassials (resembling cockleshells) that would have enabled it to kill and consume its prey efficiently. The skull measures 145 millimeters from end to end, and the lower jaw is 122 millimeters.

In less specialized members of the placental carnivore, giant rat-kangaroo and marsupial lion clans, the last two to four teeth in the upper and lower jaws are broad molars, used primarily for crushing plant material. Immediately in front of these molars are vertical shearing blades, called carnassials, that can efficiently slice through muscle, hide and sinew. Within each of these three groups of animals, however, the carnassials of the most carnivorous species are greatly enlarged, whereas the plant-processing teeth are reduced, even lost. In the mouth of a domestic cat, for instance, can be found the cheek teeth of a highly specialized carnivore.

So the relative importance of the carnassial versus the crushing teeth in an animal's jaws offers a good indication of how much flesh it devoured. In this respect, the giant rat-kangaroos resembled canids such as foxes, which are opportunistic feeders and retain significant capacity to crush. But the skull of *E. ima* featured a number of other attributes typical of carnivores. Its robust architecture, for instance, undoubtedly supported the massive neck and jaw muscles that many predators need to subdue struggling prey.

But it never evolved long canines in the lower jaw; instead its lower front incisors became daggerlike blades.

Giant rat-kangaroos were generalists, taking flesh when available but supplementing their diet with a healthy variety of vegetable matter. These renegades of the kangaroo clan terrorized the Australian continent for at least 25 million years, going extinct only sometime over the past 40,000 years.

While keeping an eye open for meat-eating kangaroos, a human intruder in Miocene Australia would have done well to avoid low-slung branches. The trees were home to another unpleasant surprise: marsupial lions. Like the giant rat-kangaroos, the four species of Miocene "lions" evolved from peaceable, plant-eating types. The most primitive species have generalized molar teeth typical of omnivores, as well as carnassial blades. In other species the crushing molars are reduced or lost, and the flesh-shearing teeth become huge.

Scientists agree that marsupial lions were indeed killers. Many consider that the most recent species, *Thylacoleo carnifex*, was the most specialized mammalian carnivore ever known: it effectively dispensed with plant-processing teeth, whereas the elaboration of its carnassials is unparalleled. It did not have big canines and must have used its long incisors to kill.

T. carnifex is also the only marsupial lion known from a complete skeleton. Many researchers have suggested that it was the size of a large wolf or leopard. Others, myself included, believe that such estimates have not accounted for the extreme robustness of the skeleton and that this frightening beast could have been as heavy as a modern lion. It was built for power, not endurance, and had tremendously muscular forelimbs. With teeth like bolt-cutters and a huge, sheathed, switchbladelike claw on the end of each semiopposable thumb, it would have been an awesome predator on any continent.

Pouched Pouncers

Undoubtedly, *T. carnifex* was adapted to take relatively large prey, probably much larger than itself. The exact purpose to which it put its thumb-claw is unclear, but one thing seems certain: once caught in the overpowering embrace of a large marsupial lion, few animals would have survived.

The kinds of marsupial lion known as *Wakaleo* were smaller, about the size of a leopard. Not designed for speed but immensely powerful, species of *Wakaleo* (and possibly *Thylacoleo*) may have specialized in aerial assault. Like the leopard, they could have launched themselves onto unsuspecting prey from trees. At the other end of the scale, at around the size of a domestic cat, *Priscileo roskellyae* may have concentrated on taking arboreal prey. Given their size and extreme predatory adaptations, I believe the larger marsupial lions most likely maintained a position at the top of the Australian food pyramid. And *T. carnifex* lived at least until 50,000 years ago—recently enough, perhaps, to have fed on humans.

On the forest floor, the marsupial wolves dominated. When Europeans arrived in Australia more than 200 years ago, they found only two marsupial families with carnivorous representatives. These were the "wolves"—only the Tasmanian Tiger remained—and a far more numerous group, the dasyurids. These mostly diminutive but pugnacious beasts are commonly measured in grams, not kilograms, and over 60 living species have been described.

Because in recent times dasyurids have clearly dominated in terms of species diversity, paleontologists had expected to find that they were also far more common than thylacinids in the distant past. We were wrong. Since 1990 seven new species of Miocene-age "wolves" have been found, bringing the total for the family to nine (including the Tasmanian Tiger). On the other hand, only one definite dasyurid has been described from Miocene deposits. A few species known from fragmentary material may also turn out to be dasyurids. Even so, the proportion of marsupial wolf to dasyurid species during the Miocene is in stark contrast to that of modern times.

The Tasmanian Tiger is the only thylacinid for which any firsthand accounts of biology and behavior are available. Most of these must be taken with a grain of salt. But the following is fairly certain: the Tasmanian Tiger was similar to most canids in that it was fully terrestrial, long-snouted and probably tended to take prey considerably smaller than itself. It differed in being relatively poorly adapted for running and probably was not a pack hunter. It further differed from the majority of canids in that its cheek teeth were adapted to a completely carnivorous diet.

In thylacinids and dasyurids the dental layout is different from that of most other flesh-eaters. These animals retain both a crushing and a vertical-slicing capacity on each individual molar. Thus, in meat-eating specialists of this type the crushing surfaces are reduced and the vertical shear is increased on each molar tooth.

Indeed, all the marsupial wolves were largely carnivorous, although the smaller, less specialized ones probably also ate insects. A number of these animals departed still further from the canid model. Some Miocene "wolves" were small compared with the Tasmanian Tiger, and one, *Wabulacinus ridei*, had a short, more catlike skull. We cannot even be sure that all Miocene-age thylacinids were terrestrial, because only fragments of the skulls and jaws are known for most. A magnificent exception is a 15-million-year-old individual recently discovered at Riversleigh; its skull and most of its skeleton are beautifully preserved. We can be reasonably certain that this animal at least lived on firm ground.

Death to Killers

Having established that Australia's large marsupial carnivores were very diverse during the Miocene period, paleontologists are now faced with this question: What happened to them? The last of the marsupial lions and giant rat-kangaroos (*T. carnifex* and *Propleopus oscillans*, respectively) died out not so long ago. In fact, they were probably around when the first Aborigines entered Australia, 50,000 or more years ago. Consequently, some scientists have maintained that it was the first humans who sounded their death knell.

Human culpability in this matter has been impossible to prove or disprove and remains a very contentious issue. No doubt the Aborigines helped to drive the Tasmanian Tiger to extinction by introducing the dingo, but their influence regarding other species is less clear-cut. These issues may never be completely resolved, but the fossil record makes one fact clear: marsupial carnivore diversity peaked by the early to middle Miocene and was already in steep decline long before humans arrived. For example, at least five marsupial wolves lived during the mid-Miocene, and two coexisted in the late Miocene, but only one was ever known to humans.

Obviously, some factor other than human influence was at work; perhaps Aborigines simply accelerated an extinction process already long established. The most likely alternative candidate is drought. From mid-Miocene times onward, Australia was subject to increasingly severe ice age conditions as well as declining rainfall and sea levels. This trend peaked over the past two million years or so, with around 20 ice ages exposing the Australian fauna to great stress. The last of these was severe, though not the worst.

Many researchers believe some combination of climate change and pressure imposed by human arrivals extinguished most of the continent's surviving larger herbivores. With their favorite meat dishes gone, the clock began to run out on Australia's marsupial predators. It is now a sad fact that of the dozens of wondrous large marsupial carnivores that have existed, not only in Australia but in the Americas as well, only our own Spotted-Tailed Quoll and Tasmanian Devil remain.

A Killer Bird?

In November 1998 Peter Murray and Dirk Megirian of the Central Australian Museum described new fossil material from an extinct, terrestrial bird called *Bullockornis planei*. This species belongs to the Australian family Dromornithidae, also called Thunder Birds, known since 1839. Dromornithids could be huge, some weighing perhaps 500 kilograms or more. But with very limited skull material preserved, little that was certain could be said about their biology. Given the paucity of material and the generally accepted view that dromornithids were closely related to predominantly plant-eating birds, most scientists were of the view that these giants were herbivores. But Murray's excellent reconstruction of *B. planei* is startling, showing a massive head possibly more than half a meter long. Furthermore, the muscle attachment sites were enormous. What did a half-ton bird with military-grade jaw muscles and a beak that could hide a football eat?

In 1991 Lawrence M. Witmer and Kenneth D. Rose convincingly argued that the massive beak and jaw musculature of *Diatryma*, an extinct bird from North America and Europe, would have constituted serious "overdesign" unless the bird was a carnivore. Following this line of reasoning, I suggested that at least some dromornithids might similarly have eaten vertebrates, killed or scavenged. If so, Thunder Birds were the largest carnivores on two legs since the demise of the meat-eating dinosaurs.

The Komodo Dragon

Claudio Ciofi

A deer nimbly picks its way down a path meandering through tall savanna grasses. It is an adult male of its species, *Cervus timorensis*, weighing some 90 kilograms (about 200 pounds). Also known as a Rusa deer, the animal knows this route well; many deer use it frequently as they move about in search of food. This Rusa's home is the Indonesian island of Komodo, a small link in a chain of islands separating the Flores Sea from the Indian Ocean. Most wildlife find survival a struggle, but for the deer on Komodo, and on a few of the nearby islands, nature is indeed quite red in tooth and claw. This deer is about to encounter a dragon.

The Komodo dragon, as befits any creature evoking a mythological beast, has many names. It is also the Komodo monitor, being a member of the monitor lizard family, Varanidae, which today has but one genus, *Varanus*. Residents of the island of Komodo may call it the *ora*. Among some on Komodo and the islands of Rinca and Flores, it is *buaja darat* (land crocodile), a name that is descriptive but inaccurate; monitors are not crocodilians. Others call it *biawak raksasa* (giant monitor), which is quite correct; it ranks as the largest of the monitor lizards, a necessary logical consequence of its standing as the biggest lizard of any kind now living on the earth. (A monitor of New Guinea, *Varanus salvadorii*, also known as the Papua monitor, may be longer than the lengthiest Komodo dragons. The lithe body and lengthy tail, however, leave it short of the thickset, powerful dragon in any reasonable assessment of size.) Within the scientific community, the dragon is *Varanus komodoensis*. And most everyone also calls it simply the Komodo.

The Komodo's Way of Life

The deer has wandered within a few meters of a robust male Komodo, about 2.5 meters (eight feet) long and weighing 45 kilograms. The first question usually asked about Komodos is: How big do they get? The largest verified specimen reached a length of 3.13 meters and was purported to weigh 166 kilograms, which may have included a substantial amount of undigested food. More typical weights for the largest wild dragons are about 70 kilograms; captives are often overfed. Although the Komodo can run briefly at speeds up to 20 kilometers per hour, its hunting strategy is based on stealth and power. It has spent hours in this spot, waiting for a deer, boar, goat or anything sizable and nutritious. Monitors can see objects as far away as 300 meters, so vision does play a role in hunting, especially as their eyes are better at picking up movement than at discerning stationary objects. Their retinas possess only cones, so they may be able to distinguish color but have poor vision in dim light. Today the tall grass obscures the deer.

Should the deer make enough noise the Komodo may hear it, despite a mention in the scientific paper first reporting its existence that dragons appeared to be deaf. Later research revealed this belief to be false, although the animal does hear only in a restricted range, probably between about 400 and 2,000 hertz. (Humans hear frequencies between 20 and 20,000 hertz). This limitation stems from varanids having but a single bone, the stapes, for transferring vibrations from the tympanic membrane to the cochlea, the structure responsible for sound perception in the inner ear. Mammals have two other bones working with the stapes to amplify sound and transmit vibrations accurately. In addition, the varanid cochlea, though the most advanced among lizards, contains far fewer receptor cells than the mammalian version. The result is an animal that is insentient to such sounds as a low-pitched voice or a high-pitched scream.

Vision and hearing are useful, but the Komodo's sense of smell is its primary food detector. Its long, yellow forked tongue samples the air, after which the two tongue tips retreat to the roof of the mouth, where they make contact with the Jacobson's organs. These chemical analyzers "smell" the deer by recognizing airborne molecules. The concentration present on the left tongue tip is higher than that sampled from the right, telling the Komodo that the deer is approaching from the left. This system, along with an undulatory walk in which the head swings from side to side, helps the dragon sense the existence and direction of odoriferous carrion from as far away as four kilometers, when the wind is right.

The Komodo makes its presence known when it is about one meter from its intended victim. The quick movement of its feet sounds like a "muffled machine gun," according to Walter Auffenberg who has contributed more to our knowledge of Komodos than any other researcher. Auffenberg lived in the field for almost a year starting in 1969 and returned for briefer study periods in 1971 and again in 1972.

The Komodo that has ambushed the deer attacks the feet first, knocking the deer off balance. When dealing with smaller prey, it may lunge straight for the neck. The basic strategy is simple: try to smash the quarry to the ground and tear it to pieces. Strong muscles driving powerful claws accomplish some of this, but the Komodo's teeth are its most dangerous weapon. They are large, curved and serrated and tear flesh with the efficiency of a plow parting soil.

Its tooth serrations harbor bits of meat from the Komodo's last meal, either fresh prey or carrion. This protein-rich residue supports large numbers of bacteria, which are currently being investigated by Putra Sastrawan, once Auffenberg's student, and his colleagues at the Udayana University in Bali and by Don Gillespie of the El Paso Zoo in Texas. They have found some 50 different bacterial strains, at least seven of which are highly septic, in the saliva.

If the deer somehow maneuvers away and escapes death at this point, chances are that its victory, and it, will nonetheless be short-lived. The infections it incurs from the Komodo bite will probably kill it within one week; its attacker, or more likely other Komodos, will then consume it. The Komodo bite is not deadly to another Komodo, however. Dragons wounded in battle with their comrades appear to be unaffected by these otherwise deadly bacteria. Gillespie is searching for antibodies in Komodo blood that may be responsible for saving them from the fate of the infected deer.

Should the deer fail to escape immediately, the Komodo will continue to rip it apart. Once convinced that its prey is incapacitated, the dragon may break off its offensive for a brief rest. Its victim is now badly injured and in shock. The Komodo suddenly launches the coup de grâce, a belly attack. The deer quickly bleeds to death, and the Komodo begins to feed.

The muscles of the Komodo's jaws and throat allow it to swallow huge chunks of meat with astonishing rapidity: Auffenberg once observed a female who weighed no more than 50 kilograms consume a 31-kilogram boar in 17 minutes. Several movable joints, such as the intramandibular hinge that opens the lower jaw unusually wide, help in the bolting. The stomach expands easily, enabling an adult to consume up to 80 percent of its own body weight in a single meal, which most likely explains some exaggerated claims for immense weights in captured individuals.

Large mammalian carnivores, such as lions, tend to leave 25 to 30 percent of their kill unconsumed, declining the intestines, hide, skeleton and hooves. Komodos eat much more efficiently, forsaking only about 12 percent of the prey. They eat bones, hooves and swaths of hide. They also eat intestines, but only after swinging them vigorously to scatter their contents. This behavior removes feces from the meal. Because large Komodos cannibalize young ones, the latter often roll in fecal material, thereby assuming a scent that their bigger brethren are programmed to avoid consuming.

More Komodos, attracted by the aromas, arrive and join in the feeding. Although males tend to grow larger and bulkier than females, no obvious morphological differences mark the sexes. One subtle clue does exist: a slight difference in the arrangement of scales just in front of the cloaca, the cavity housing the genitalia in both sexes. Sexing Komodos remains a challenge to researchers; the dragons themselves appear to have little trouble figuring out who is who. With a group assembled around the carrion, the opportunity for courtship arrives.

Most mating occurs between May and August. Dominant males can become embroiled in ritual combat in their quest for females. Using their tails for support, they wrestle in upright postures, grabbing each other with their forelegs as they attempt to throw the opponent to the ground. Blood is usually drawn, and the loser either runs or remains prone and motionless.

The victorious wrestler initiates courtship by flicking his tongue on a female's snout and then over her body. The temple and the fold between the torso and the rear leg are favorite spots. Stimulation is both tactile and chemical, through skin gland secretions. Before copulation can occur, the male must evert a pair of hemipenes located within his cloaca, at the base of the tail. The male then crawls on the back of his partner and inserts one of the two hemipenes, depending on his position relative to the female's tail, into her cloaca.

The female Komodo will lay her eggs in September. The delay in laying may serve to help the clutch avoid the brutally hot months of the dry season. In addition, unfertilized eggs may have a second chance with a subsequent mating. The female lays in depressions dug on hill slopes or within the pilfered nests of Megapode birds. These chicken-size land dwellers make heaps of earth mixed with twigs that may reach a meter in height and three meters across. While the eggs are incubating, females may lie on the nests, protecting their future offspring. No evidence exists, however, for parental care of newly hatched Komodos.

The hatchlings weigh less than 100 grams and average only 40 centimeters in length. Their early years are precarious, and they often fall victim to predators, including their fellow Komodos. They feed on a diverse diet of insects, small lizards, snakes and birds. Should they live five years, they can weigh 25 kilograms and stretch two meters long. By this time, they have moved on to bigger prey, such as rodents, monkeys, goats, wild boars and the most popular Komodo food, deer. Slow growth continues throughout their lives, which may last more than 30 years. The largest Komodos, three meters and 70 kilograms of bone, teeth and sinew, rule their tiny island kingdoms.

The Komodo's Past

Komodos, as members of the class Reptilia, do have a relationship with dinosaurs, but they are not descended from them, as is sometimes believed. Rather Komodos and dinosaurs share a common ancestor. Both monitor lizards and dinosaurs belong to the subclass Diapsida, or "two-arched reptiles," characterized by the presence of two openings in the temporal region of the skull. The earliest fossils from this group date back to the late Carboniferous period, some 300 million years ago.

Two distinct lineages arose from those early representatives. One is Archosauria, which included dinosaurs. The ancestor of monitor lizards, in contrast, stemmed from primitive Lepidosauria at the end of the Paleozoic era, about 250 million years ago. Whereas some dinosaurs evolved upright stances, the monitor lineage retained a sprawling posture and developed powerful forelimbs for locomotion. During the Cretaceous, and starting 100 million years ago, species related to present-day varanids appeared in central Asia. Some of these were large marine lizards that vanished with the dinosaurs, about 65 million years ago. Others were terrestrial forms, up to three meters in length, that preyed on smaller animals and probably raided dinosaur nests. About 50 million years ago, during the Eocene, these species dispersed throughout Europe and south Asia and even into North America.

Varanids reached Australia by about 15 million years ago, thanks to a collision between the Australian landmass and southeast Asia. Numerous small varanid species, known as pygmy monitors, quickly colonized Australia, filling multiple ecological niches. More than two million years later a second lineage differentiated and spread throughout Australia and the Indonesian archipelago, which was at the time far closer to Australia than it is today, because much of the continental shelf was above water. *V. komodoensis* is a member of that lineage, having differentiated from it about four million years ago.

The Indo-Australian varanids could take advantage of their unique faunal environment. Islands simply have fewer resources than large landmasses do. Because reptilian predators can subsist on much lower total energy requirements than mammals can, a reptile will have the advantage in the race for top predator status under these conditions.

In such a setting, reptiles can also evolve to huge size, an advantage for hunting. A varanid called *Megalania prisca*, extinct for around 25,000 years, may have reached a length of six meters and a weight of 600 kilograms; the late extinction date means that humans may have encountered this monster. Komodos adopted a more moderate giantism. Reasons for the Komodo's current restricted home range—the smallest of any large predator—are the subject of debate and study. Various researchers subscribe to alternative routes that the dragons' ancestors may have taken to their present locale of Komodo, Flores, Rinca, Gili Motang and Gili Dasami.

Komodo has a different paleogeography from its neighbors. According to worldwide sea-level changes over the past 80,000 years and bathymetric data of the study area, Flores and Rinca were joined until 10,000 years ago.

Gili Motang was connected several times to their combined landmass. Komodo was long isolated but appears to have joined its eastern neighbors about 20,000 years ago, during the last glacial maximum. That association may have lasted 4,000 years.

The World Discovers a Dragon

The West was unaware of the Komodo until 1910, when Lieutenant van Steyn van Hensbroek of the Dutch colonial administration heard local stories about a "land crocodile." Members of a Dutch pearling fleet also told him yarns about creatures six or even seven meters long. Van Hensbroek eventually found and killed a Komodo measuring a more realistic 2.1 meters and sent a photograph and the skin to Peter A. Ouwens, director of the Zoological Museum and Botanical Gardens at Bogor, Java.

Ouwens recruited a collector, who killed two Komodos, supposedly measuring 3.1 and 2.35 meters, and captured two young, each just under one meter. On examination of these specimens, Ouwens realized that the Komodo was in fact a monitor lizard. In the 1912 paper in which Ouwens introduced the Komodo to the rest of the world, he wrote simply that van Hensbroek "had received information . . . [that] on the island of Komodo occurred a *Varanus* species of an unusual size." Ouwens ended the paper by suggesting the creature be given the name *V. komodoensis*.

Understanding the Komodo to be rare and magnificent, local rulers and the Dutch colonial government instituted protection plans as early as 1915. After World War I, a Berlin Zoological Museum expedition roused worldwide interest in the animal. In 1926 W. Douglas Burden of the American Museum of Natural History undertook a well-equipped outing to Komodo, capturing 27 dragons and describing anatomical features based on examinations of some 70 individuals.

The Komodo's Future

More than 15 expeditions followed Burden's, but it was Auffenberg who performed the most comprehensive field study, looking at everything from behavior and diet to demographics and the botanical features of their territory. Auffenberg determined that the Komodo is, in fact, rare. Recent estimates suggest that fewer than 3,500 dragons live within the boundaries of Komodo Island National Park, which consists of the islands of Komodo (1,700 individuals), Rinca (1,300), Gili Motang (100) and Padar (none since the late 1970s), and some 30 other islets. A census on Gili Dasami has never been done. About another 2,000 Komodos may live in regions of the island of Flores. The Komodo is now officially considered a "vulnerable" species, according to the World Conservation Union; it is also protected under the Convention on International Trade in Endangered Species of Wild Fauna and Flora.

The Komodo dragon has faced major challenges that threaten its survival in part of the national park and on Flores. The disappearance of dragons on Padar probably stems from poaching of their primary prey, deer. Policing this rugged and sometimes inaccessible habitat is difficult. Nevertheless, a trend toward less poaching overall on Padar has moved officials to discuss a reintroduction program.

Padar covers an area of only about 20 square kilometers and supports no more than 600 deer, in turn limiting the number of Komodos. Consequently, genetic diversity, as insurance against inbreeding, would be highly desirable among a new, small Komodo population.

Komodos on Flores face the twin threats of prey depletion and habitat encroachment by humans. New settlers slash and burn the monsoon forest, and Komodo dragons are among the first species to disappear. Over the past 20 years, habitat loss has caused the species to vanish from an area stretching for 150 kilometers along Flores's northwest coast. Populations on the north and west coasts are also threatened by deforestation and indirectly through deer hunting.

The fortunes of the Komodo dragon are inexorably linked with those of numerous other species of fauna and flora, and measures to protect this giant lizard must take into account the entirety of its natural habitat. For example, although central Flores is inhospitable to dragons, the southern and eastern regions of the island may harbor scattered populations, still unknown to researchers, that could act as "umbrellas" to protect the ecosystem as a whole. The charismatic dragon already draws some 18,000 visitors a year to the area, and patches of forest containing Komodos could be the cornerstone of an economically viable protection plan for the entire habitat, based on ecotourism.

The fate of the world's few thousand Komodos, living out their lives in a tiny corner of the earth, is probably now in human hands. Policy decisions, as in so many wildlife conservation issues, will be as much aesthetic as scientific or economic. We can choose to create a homogeneous world of stultifying sameness. Or we can choose to maintain a remnant of the mystery that provoked medieval cartographers to mark the unexplored territories of their maps with the exhilarating warning, "Here there be dragons."

The Terror Birds of South America

Larry G. Marshall

It is a summer day on the pampas of central Argentina some five million years ago. A herd of small, horselike mammals are grazing peacefully in the warm sun. None of the animals is aware of the vigilant creature standing 50 meters away in the tall grass. Most of the watcher's trim, feathered body is concealed by the vegetation. Its eyes, set on the sides of a disproportionately large head perched on a long and powerful neck, are fixed on the herd. The head moves from side to side in short, rapid jerks, permitting a fix on the prey without the aid of stereoscopic vision.

Soon the head drops to the level of the grass, and the creature moves forward a few meters, then raises its head again to renew the surveillance. At a distance of 30 meters, the animal is almost ready to attack. In preparation, it lowers its head to a large rock close to its feet, rubbing its deep beak there to sharpen the bladelike edges.

Now the terror bristles its feathers and springs. Propelled by its two long, muscular legs, it dashes toward the herd. Within seconds it is moving at 70 kilometers per hour. Its small wings, useless for flight, are extended to the sides in aid of balance and maneuverability.

The herd, stricken with fright, bolts in disarray as the predator bears down. The attacker fixes its attention on an old male lagging behind the fleeing animals and quickly gains on it. Although the old male runs desperately, the attacker is soon at its side. With a stunning sideswipe of its powerful left foot, it knocks the prey off balance, seizes it in its massive beak and, with swinging motions of its head, beats it on the ground until it is unconscious. Now the attacker can swallow the limp body whole—an easy feat, given the creature's meter-long head and half-meter gape. Content, the gorged predator returns to its round nest of twigs in the grass nearby and resumes the incubation of two eggs the size of basketballs.

Meet the terror birds, the most spectacular and formidable group of flightless, flesh-eating birds that ever lived. They are all extinct now, but they were once to the land what sharks are to the seas: engines of destruction and awesome eating machines. In their time, from 62 million years to about 2.5 million years ago, they became the dominant carnivores of South America. The story of their rise and decline is my subject here.

The terror birds are members of a group ornithologists call phorusrhacoids. The first phorusrhacoid to be described scientifically—in 1887 by the Argentine paleontologist Florentino Ameghino—was a fossil that he named *Phorusrhacos longissimus*. The fossil came from the Santa Cruz Formation in Patagonia, the southernmost region of Argentina; the formation is about 17 million years old.

Ameghino and other researchers reconstructed the appearance of the birds from their fossil remains and their behavior from what creatures that might be living relatives do. The investigators initially interpreted the flesh-eating habits of the phorusrhacoids as an indication that they were related to modern eagles and hawks. Not all paleontologists agreed, and the issue was debated over the next 12 years. Charles W. Andrews of the British Museum resolved the controversy in 1899, concluding that among all living and extinct groups, the phorusrhacoids were most closely related to the South American seriema birds, which could also be regarded as the structural ancestors of the phorusrhacoids. Seriemas live today in the grasslands of northern Argentina, eastern Bolivia, Paraguay and central and eastern Brazil. Seriemas and phorusrhacoids are classified as members of the order Gruiformes, which includes cranes and rails and their kin.

There are two living seirema species, the red-legged seriema (*Cariama Cristata*) and the black-legged, or Burmeister's, seriema (*Chunga burmeisteri*). These birds reach a height of 0.7 meter. They are light-bodied, long-legged and long-necked. Their wings are small relative to their body, and the birds resort to spurts of short-distance flight only when pressed. They are excellent runners, able to attain speeds in excess of 60 kilometers per hour. Seriemas build twig nests, four to six meters above the ground, in low trees. The young, usually two, mature in about two weeks, whereupon they leave the nest to live and hunt in the nearby grasslands. Like most carnivorous animals, seriemas are territorial. Their call has been described as eerie and piercing.

Like the phorusrhacoids, seriemas are carnivorous. They eat insects, reptiles, small mammals and other birds. Under favorable conditions, they will attack larger game. They seize their prey in their beaks and beat the animal on the ground until it is limp enough to be swallowed whole. This feeding strategy is also practiced today by the roadrunner (*Geococcyx californianus*) of the southwestern U.S. and the secretary bird (*Sagittarius serpentarius*) of Africa.

Seriemas are placed in the family Cariamidae, which now is restricted to South America. About 10 fossil species have been found there, the oldest being from the middle Paleocene epoch (some 62 million years ago) of Brazil. Relatives of this group are represented by two fossil families: the Bathornithidae, which appear in beds 40 to 20 million years old in North America, and the Idiornithidae, found in certain European rock formations 40 to 30 million years old. Some workers believe these families are so closely related that they should all be grouped in the family Cariamidae.

Most of the terror birds were considerably larger than their living relatives. The creatures ranged in height from one to three meters (just shy of 10 feet). The earliest known members are virtually as specialized as the latest, indicating that they originated before their first appearance in the fossil record.

About a dozen genera and 25 species of terror birds have been recognized. The relation among them is still not clear. They were classified in 1960 by Bryan Patterson and Jorge L. Kraglievich. This classification ordered the terror birds in three families that, in comparison to families of mammals, include animals of medium, large and gigantic size.

The gigantic forms are members of the family Brontornithidae. Fossils of this family have been found in beds ranging in age from 27 to 17 million years. A heavy, ponderous build characterized the birds; the leg bones were fairly short, the beaks massive. This evidence suggests that the birds were cumbersome runners, slower afoot than the members of the other two families.

Next comes the family Phorusrhacidae. Its members ranged between two and three meters in height. Fossils have been found in rocks ranging in age from 27 million to three million years. The third family, Psilopteridae, comprised quite small members; most of them stood no more than one meter in height. Their known fossils range from 62 million to two million years in age. Within this family is the oldest known phorusrhacoid, *Paleopsilopterus*, found in Brazil. Members of these last two families were lightly built, swift runners. They were the ones that became the dominant running carnivores of their time, and they held that status for millions of years.

The fact that phorusrhacoids came in several sizes indicates that the adults were capable of preying on a wide variety of animals, from rodents to large herbivores. Although some of the adult herbivores were as big as some adult phorusrhacoids, the birds could easily have preyed on the young ones. Phorusrhacoids newly out of the nest would have had different food needs because they were smaller; they probably hunted rodents and other small vertebrates, much as their living seriema relatives still do.

During much of the age of mammals (the past 66 million years), phorusrhacoids thus occupied the role of fleet-footed carnivores in South America. They were able to assume this role by giving up the greatest virtue of being a bird—the power of flight. The door to dominance as carnivores opened to the phorusrhacoids when their predecessors in that role—the small, bipedal dinosaurs known as coelurosaurs—disappeared in the dinosaur extinction 66 million years ago. Paleobiologists call such a transition an evolutionary relay.

The body forms of the terror birds and the coelurosaurs were quite similar: trim, elongated bodies; long, powerful hind limbs; long necks; large heads. Many coelurosaurs had reduced anterior limbs, indicating that the animals captured, killed and processed prey primarily with the hind limbs and mouth, as the phorusrhacoids did. Coelurosaurs apparently used their long tail as a balance while running; phorusrhacoids probably used their reduced wings for the same purpose. Different strategies and appendages were thus employed to serve the same functional purpose.

Terror birds and their relatives are also known outside South America. One can begin to see why a group of large, flightless birds rose to the top of the food pyramid and why they finally lost that position. The answer lies in the historical development of the terrestrial fauna of South America. Recall that for most of the past 66 million years South America was, as Australia is today, an island continent. As a consequence of the groups that inhabited each continent 66 million years ago, the role of terrestrial mammal carnivores was filled in South America by marsupials and the role of large herbivores by placentals. This marsupial-placental combination was unique among continental faunas; both roles were filled by marsupials in Australia and by placentals in North America, Europe and Asia.

The group of South American marsupials that evolved to fill the place that placental dogs and cats eventually held on the northern continents is called borhyaenoid. Its doglike members are further grouped into three families. They ranged in size from that of a skunk to that of a bear. One specialized family, the thylacosmilids, had characteristics similar to those of the placental saber-toothed cats. It is particularly significant that all these animals were relatively short-legged and that none showed marked adaptation to running. These were the mammal occupants of the carnivore niche in South America.

From about 27 million to 2.5 million years ago, the fossil record shows a protracted decrease in the size and diversity of the doglike borhyaenoids and a concurrent increase in the size and diversity of the phorusrhacoids. Consequently, by about five million years ago, phorusrhacoids had completely replaced the large carnivorous borhyaenoids on the savannas of South America. (The smaller ones, which were not competitive with the terror birds anyway, also became extinct before the Panamanian land bridge appeared.) This transition demonstrates another relay in the evolutionary history of the phorusrhacoids whereby they successfully replaced their marsupial counterparts, the borhyaenoids. Just why the phorusrhacoids were able to do so is unclear, but their superior running ability would certainly have been an advantage for capturing prey in the savanna environments that first came into prominence about 27 million years ago.

After the emergence of the Panamanian land bridge, placental dogs and cats of the families Canidae and Felidae dispersed into South America from North America. Because all the large marsupial carnivores of South America were by then long extinct, the only competition the dogs and cats had was the phorusrhacoids. It proved to be a losing battle for the birds.

The terror birds thus flourished in the absence of advanced placental carnivores, which have repeatedly shown themselves to be better competitors. The marsupial borhyaenoids and placental creodonts were, in essence and in comparison with the terror birds, second rate.

Although plausible, this argument is speculative. One cannot identify with certainty a single factor that explains the extinction of any group of animals now found only as fossils. In the case of the terror birds, their disappearance on two occasions in time correlates directly with the appearance of advanced placental carnivores. Were the advanced placentals more intelligent than the terror birds and so better adapted to capturing the prey that the birds had had to themselves? Did the fact that they had four legs give them an advantage over the two-legged phorusrhacoids in speed or agility? Did the placentals eat the phorusrhacoids' eggs, which were readily accessible in ground nests because of the birds' large size? Did the placentals prey on the vulnerable hatchlings?

It is intriguing to think what might happen if all big carnivorous mammals were suddenly to vanish from South America. Would the seriemas again give rise to a group of giant flesh-eating birds that would rule the savannas as did the phorusrhacoids and their bygone allies?

Shrews

Oliver P. Pearson

To a biologist a shrew is a vexatious, scolding, turbulent mammal (order of insect eaters) that has rarely submitted to taming since long before Shakespeare. Because shrews are extremely tiny, live in burrows and seldom interfere with man's interests and activities, even their unruly personality has failed to bring them to people's attention. Their obscurity is certainly not due to scarcity; in the eastern third of the U.S., for instance, probably the most abundant mammal is a certain species of shrew. More than 30 species are found in North America, and the animal makes its home on every major continent except Australia.

The American shrews resemble small moles (to which they are related) and earless mice (no relation). They have long, pointed noses, tiny eyes and velvety fur. Some of them are astonishingly small, weighing less than a penny. The shrew's plushlike fur is well-suited to its life in burrows, for no matter in which direction the animal goes its hair does not muss. Its eyes are almost hidden in its fur. It has little need of them, as a matter of fact, for it spends practically all its time in dark tunnels or thick vegetation, where it must rely primarily on hearing (which is good despite its negligible ear lobes), on smell and on tactile stimuli, received by sensitive whiskers near the tip of its nose. Shrews move about constantly, quivering their nose nervously and sometimes keeping up a faint, thin, high-pitched twittering. Some individuals of the common short-tailed species when cornered throw their head back, open their mouth and utter a long, shrill chatter.

The shrew's living requirements are moisture, cover and a supply of small invertebrates and seeds for food. It is so small that it can make its home in leafmold, grass, sphagnum or marsh plants, where it digs a tunnel or makes a leafy nest. It lives in such diverse environments as forests, meadows, bogs and salt marshes. Shrews like deciduous forests that are rich in leaf litter and rotting logs. Rolling aside one of these logs frequently discloses a shrew tunnel or nest. The little animal finds plenty of worms, grubs and insects within its tunnel and runways, but it must also scurry out at frequent intervals for nuts and seeds to fill out its diet.

So elusive are shrews that even experienced investigators seldom see them in the wild. About the only way to get a look at them is to capture them in mousetraps. Traps placed at their runways and other favorable spots will catch a fair number, but the captures give no true idea of the actual shrew population, for some of the species are trap-shy. I trapped for many years in suburban Philadelphia, using all sorts of ingenious (I thought) traps, without catching a single shrew of the species *Cryptotis parva*. Yet I discovered that barn owls were catching these little shrews in the same area. The first specimens of Cryptotis ever found in Canada were discovered (still alive) in the stomach of a milk snake. Some years ago Marcus Lyon, an Indiana mammalogist, reported that in many years of trapping he had captured only one Cryptotis in that state. But on Christmas Day, 1949, a hawk shot in Wayne County, Indiana, had in its digestive system the remains of 27 Cryptotis!

Shrews are noted for their nervous restlessness, and there is good reason for this. As the smallest living mammals, they have the highest rate of metabolism, for the "living" rate of animal tissue goes up as animal size goes down. One of the smaller species of shrew, *Sorex cinereus*, metabolizes four times as fast as a mouse per gram of tissue. To support their high rate of metabolism shrews must devour enormous quantities of food. To be impressed by the appetite of the shrew, you need only capture one and try to keep it fed. You will soon weary of any attempt to catch enough worms, grubs and insects to satiate it and will have to resort to teaching the shrew to eat dog food and ground meat. One man found that his 3.6-gram Sorex consumed in eight days more than 93 grams of food—earthworms, rolled oats, mice, sow bugs, grasshoppers, snails, fish and whatnot. The shrew ate on the average 3.3 times its own weight per day. C. Hart Merriam, a pioneer American mammalogist, once confined three Sorex under a glass tumbler. Two of them promptly attacked and devoured the third. Eight hours later only a single shrew, with slightly bulging stomach, remained. The tiny cannibal had transmuted both of its companions into a small heap of droppings and a few calories of wasted heat.

Such voracity is ordinarily directed toward small invertebrates, but *Blarina*, one of the larger shrews, frequently eats mice two or three times its own size. It readily kills big meadow mice. A. F. Shull, once found a Blarina nest made exclusively of mouse hair. Beside this nest were two freshly killed meadow mice and the body of a third half-eaten, and nearby lay several handfuls of hair in which were mixed the legs and tails of about 20 more. Shull was prompted to attempt an estimate of the predatory impact of shrews on meadow mice. Assuming that mice made up about 40 percent of the diet of these shrews, he calculated that each shrew would consume about eight mice per month. Figuring a population of four shrews per acre, on a 100-acre farm the total number of mice consumed by shrews in a year would be 38,400!

Such statistics are frequently full of pitfalls, but in this case may not be far misleading. Recent studies by Robert Eadie support Shull's conclusions, if not his statistics. Eadie, like other collectors, had frequently been humiliated by having shrews invade his mousetraps and escape, leaving their droppings on the floor. Making the best of this frustrating discovery, he left large numbers of paper squares about to collect shrew droppings, usually hard to find. After analyzing the content of the droppings over many seasons, he concluded that Blarina shrews catch large numbers of meadow mice, even in years when the mice are scarce, and probably are important regulators of the meadow-mouse population.

Blarina possesses a venom that helps to subdue its prey. As far as we know it is the only mammal with a poisonous bite, although some European and African shrews are suspected also. Its venom is powerful enough to kill a human being if injected into the bloodstream, but the animal lacks an effective injection mechanism. Unlike poisonous snakes, it has no hollow fangs; the poison seeps into its bite from a groove between its long lower incisor teeth. This mechanism seems to be adequate to kill a mouse but not an animal as large as a rabbit.

In 1889 a New England naturalist, C. J. Maynard, reported the effects of a bite on the hand while he was trying to capture a Blarina shrew. His skin was slightly punctured in a number of places. Within 30 seconds he felt a burning sensation, and soon afterward shooting pains ran up his arm. The pain and swelling reached a maximum in about an hour. He could not use his hand without great suffering for three days and felt considerable discomfort for more than a week afterward. His report of the incident was published in an obscure journal at a time when Pasteur's work on rabies and "microbes" held the center of attention. Sophisticated scientists of the day attributed Maynard's symptoms to microbes and dismissed the idea that the shrew had injected a poison.

Actual proof of the toxicity of shrews came. In the course of a joint reconnaissance of the microscopic anatomy of the common short-tailed shrew of the eastern U.S., George Wislocki pointed out to me an unusual group of cells in the microscopic tubules of the shrew's submaxillary salivary glands (below the lower jaw). It seemed worthwhile to test these cells as a possible source of venom. Extracts from the glands, made by grinding them in salt solution, were injected into mice. They proved highly lethal to the mice. In fact, only one-200th of the submaxillary extract from a single Blarina will kill a white mouse within a few minutes when injected into its bloodstream. The gross symptoms are labored breathing, protruding eyes and convulsions.

Despite their abundance shrews are not exceptionally prolific. A female of the short-tailed species, for example, has no more than two or three litters in a year—one or two in the spring and, if she survives, another in the summer. The average size of the litter is around five. Females born in the spring mature rapidly and soon can reproduce, but those born in the autumn do not reproduce until the following spring.

In captivity shrews can live more than two and a half years. In the wild, however, they age rapidly and encounter so many hazards that few live to be a year old. Among a sample of short-tailed shrews marked and released in the summer only a little more than 6 percent survived until the following summer. And since a large number of young perish even before they are weaned or are old enough to be caught, marked and released, it is safe to say that the life expectancy of a shrew at birth is not more than a few months. Survival of the species is left to the immature, inexperienced generation that lives over the winter—a situation reminiscent of many species of insects, whose adults perish by the end of the season and entrust the preservation of the race entirely to larvae and pupae.

There has been some dispute about whether shrews die more commonly of old age or of violence. The facts that their tissues live at a rapid rate and that the uneaten bodies of dead shrews are sometimes found lying about in the woods and fields (more often than the bodies of dead mice) seem to suggest that many die of old age. However, nearly all the shrew carcasses I have found have shown clear evidence, beneath their fur, of a violent death. A record of the age distributions of samples of the shrew population trapped in various months of the year shows a high attrition among all groups, young and old, as the year goes on.

The simplest explanation of the carcasses found in nature is that shrews are unpalatable to many predators. They have powerful scent glands in the skin. The short-tailed shrew gives forth a particularly offensive odor from oily skin glands on each side and along the midline of the belly. The odor seems to become stronger when the shrews are excited or angry. It renders shrews less palatable to foxes, cats, weasels and probably to many other animals that prey on shrews. It offers no protection against hawks or owls, however, because they have no sense of smell.

In the Blarina genus of shrews, females in the breeding stage have less well-developed scent glands than the rest of the population. This is rather surprising, because for the sake of the population those females are the most in need of protection against predators. Probably the answer is that the scent glands also serve another purpose—a social one. A wandering male shrew encountering an unscented runway or tunnel may assume that it is either vacant or occupied by a breeding female, and so not hesitate to enter. On the other hand, males are heavily scented during the breeding season. The scent, rubbing off on their burrows, may serve both to attract breeding females and as a keep-out sign to others. Birds use song in the same way, and male dogs stake territorial claims by scenting tree trunks and other objects.

Konrad Lorenz, in his charming book about animals, *King Solomon's Ring*, tells of a family of captive European water shrews that memorized a particular pathway in their cage. They scurried unhesitatingly along this path, like a locomotive on its track, until some minor rearrangement was made in the pathway, whereupon the shrews were thrown into confusion. At one point on the route the shrews were accustomed to jump onto a pile of small stones and then off the other side. When the stones were removed, shrews coming along the path jumped into the air at the appropriate place but landed on the floor of the cage with a disconcerting bump. Then, despite the fact that their vision is good enough to see obstructions, the shrews made their way back onto the accustomed pathway and with restored confidence repeated the pointless leap. Although this behavior appears to demonstrate gross stupidity, it should be pointed out in the shrew's favor that its remarkable path-memory normally releases its other senses for more important activities. Its substitution of habit for reason may seem unenlightened to us, but the abundance and success of shrews throughout the world are strong evidence that in the shrew's case the substitution is effective.

Snakebite

Sherman A. Minton, Jr.

Herpetologists insist that snakes are friends of mankind, but most of us are glad to steer clear of these friends. After due credit has been given them for their services in consuming insects and vermin, and for their fascinating virtues as pets and show animals, we must still take a cold view of their venomous aspects. Just how dangerous are the poison snakes? Prudence and self-protection make it worthwhile to look into this question, biologically and statistically.

It is a simple matter to measure the toxicity of a snake's venom by injecting carefully weighed doses into an experimental animal. But extrapolation from these tests to estimate the fatal dose for man may be misleading, if not dangerous. A favorite animal for such tests used to be the pigeon, which is highly sensitive to snake poison. Pigeon assays probably are the basis for the statement in a well-known textbook on tropical medicine that the fatal human dose of the venom of the North American copperhead is about 25 milligrams (dried). Since an average adult copperhead can deliver 50 milligrams of venom at a bite, many persons bitten by copperheads should, on this figuring, be candidates for the undertaker. Actually the copperhead bite is rarely fatal, even when the bitee gets no medical treatment.

At the other extreme from the pigeon is the white mouse, which has replaced the pigeon as a favorite assay animal because it is so much more inexpensive and convenient to handle. The white mouse has amazing resistance to snake venoms, especially those of the pit vipers. Ounce for ounce, it is more than 12 times as resistant as the pigeon to a rattlesnake's poison. Now on the basis of mouse tests of the venom of the fer-de-lance, one might calculate the lethal dose of this tropical snake's poison for a 150-pound man as about 1,550 milligrams. But even a very large fer-de-lance delivers only about 300 milligrams at a bite. Thus we might conclude that one of the most dangerous snakes of tropical America should be unable to inflict fatal injury to an adult human being. Its hundreds of victims no doubt succumbed to fright!

Snakes, like other animals, vary in individual makeup and condition, and it is well to remember that this applies to the toxicity of their venom. Testing samples collected from two female timber rattlesnakes on the same day in the same place, I found one approximately five times more toxic than the other. This sort of variation can lead to real discrepancies, especially in the case of rare species when an investigator's entire sample may have come from only one or two animals.

What snakes have the most toxic venom? After giving due weight to all pertinent factors, we would put at the top of the list the poisons of the tiger snake of Australia (*Notechis scutatus*), the blue krait of India (*Bungarus candidus coeruleus*) and the krait of Formosa (*B. multicinctus*). Their venoms are lethal to the mouse in a dose of only two- to five-thousandths of a milligram: they are almost 50 times as toxic as potassium cyanide. There are several less studied snakes whose venoms are probably of the same order of toxicity. Among them are the other species of kraits, the mambas of Africa (related to the cobras), the taipan of Australia and some of the sea snakes. The venoms of all these snakes are toxic to the nervous system, producing paralytic symptoms and death from depression of respiration and the heartbeat.

Extremely toxic venom does not in itself make a snake dangerous to man. It becomes a menace only if the snake makes the poison in sufficient quantity, has an effective means of delivering it and is skillful and aggressive in doing so. Many snakes with very powerful venoms have only small amounts of this secretion available for a bite. The tiger snake, for example, delivers less than a fifth as much venom (about 40 milligrams dry weight) as some of the large rattlesnakes. Other poisonous snakes are handicapped by having short fangs or an otherwise inefficient biting apparatus. On the other hand, a large snake with comparatively weak venom may be very dangerous.

Aggressiveness in serpents is extremely difficult to evaluate and requires intimate knowledge of the species. There is enormous individual variation.

Many snakes are aggressive only under certain conditions or at particular times. An experienced herpetologist told me of his first experience with

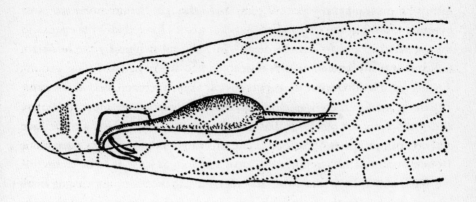

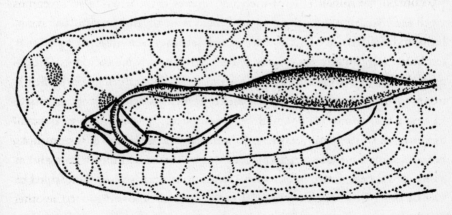

Fangs and poison gland of the Indian cobra (top) are compared with those of the western diamondback rattlesnake (bottom). The poison apparatus of the cobra is smaller, but its venom is more toxic.

kraits in Java: "When the natives brought them in during the day, they were just as limp as so many pieces of rope, and I thought the krait was the most overrated snake in the East Indies. But that night I looked into the cage, and they were alert and full of the devil." In India most bites by the krait are inflicted at night, sometimes upon sleeping persons.

Among the other large and aggressive snakes, the mambas (*Dendroaspis angusticeps* and *D. polylepis*) of the African forests have great quickness and a fondness for climbing that permits them to inflict bites on the face and other upper parts of the body. The Australian taipan reaches a length of 10 feet or more and is reported to have a venom at least as toxic as that of the tiger snake. It is said to attack without provocation, and it has been known to bite effectively through a boot and a heavy sock.

Occasionally a dangerous species may be underestimated by authorities who have had only limited experience with it. The great American herpetologist John E. Holbrook considered the coral snake of the southern U.S. (*Micrurus fulvius*) almost innocuous, and this opinion was prevalent among naturalists for many years. Our information now indicates, however, that coral snake bites are fatal in at least 20 percent of the cases.

One of the most deadly of the small snakes is the saw-scaled or carpet viper (*Echis carinatus*), found in arid regions of the subtropical belt from India to West Africa. This evil-tempered serpent's venom is so toxic that fatalities from its bite have exceeded 80 percent in some localities. A herpetologist who served with the U.S. Air Force in India during World War II returned with two of these reptiles and told me: "When I saw the first one, I knew it was some sort of small viper and guessed it was no more dangerous than our little pigmy rattlers. I didn't have a stick, so I just pinned its head with my fountain pen and grabbed it by the neck. I didn't know what a chance I'd taken until later, when I heard that a native had died from the bite of one of these snakes just 10 inches long. The one I picked up was more than twice that size."

A snake in the wild is not, of course, dangerous to the human population, and so some of the most deadly species are not much of a menace. For example, the taipan, a truly formidable reptile, kills few persons, for it has a rather limited range in thinly settled northern Australia. The mambas, the bushmaster and to a lesser extent the king cobra are primarily forest snakes whose contacts with human beings are rare. It is the snakes that thrive in heavily populated areas of the world that account for most of the deaths from snakebite. In some countries they are a fairly important public health problem.

How much of a problem they are depends in part on the customs and economy of the region. In the midwestern farm belt of the U.S. snakebites are rare, because mechanized agriculture creates conditions unfavorable for snakes and minimizes the farmers' exposure. In tropical and subtropical plantations, where hand labor predominates, the danger is much greater. Brush and other rubbish piled in the fields provides habitats for snakes, and some types of agriculture foster rats and other small rodents, which are a favorite food of many poisonous snakes. Irrigation seems to have been followed by increase of some kinds of rattlesnakes in the U.S. Southwest. Houses of thatch or loosely laid stone may provide refuges for snakes. And people who go about in bare legs or feet are vulnerable to short-fanged snakes such as the kraits. People who sleep on mats spread on the ground run a greater risk than those who use beds or hammocks.

Religious protection of snakes (e.g., in parts of India, among the Hopi in the U.S. and in rattlesnake cults of southern mountaineers) increases the snakebite toll, and so too does the profession of snake charming. When religious zealots or entertainers handle poisonous snakes, the risk is very great. More often than not, however, the snakes are "milked" before handling, or their fangs are broken, or their mouths are sewed shut or otherwise mutilated. Some snake charmers actually handle only nonpoisonous reptiles, although they may exhibit poisonous ones.

Reliable information on the number of snakebites or snakebite deaths is at best difficult to obtain: for a region such as tropical Africa there is not enough information even for an intelligent guess. But the World Health Organization has collected approximate figures for a number of areas of the world. These indicate that southeastern Asia has the highest death rates from snakebite. The most dangerous area seems to be in the Irrawaddy and Chindwin valleys of Burma.

Next to Southeast Asia the second most dangerous area is tropical America, where the snakebite mortality is about 3,000 to 4,000 annually.

North America, according to the World Health Organization study, has some 300 to 400 snakebite deaths per year. Only a minority of these occur in the U.S. It is likely that the average number of deaths from snakebites in the U.S. is less than 50 a year. The low mortality rate can be credited in part to the fact that snakebite victims usually get prompt and lavish medical care.

It is plain that mankind has reason to fear snakes, and rather surprising that more study has not been given to their venomous powers and contacts with the human population. Further research would make these dangerous enemies less formidable.

Suggested Reading

Attenborough, David. *Zoo Quest for a Dragon*. Lutterworth Press, 1957. London: Oxford University Press, 1986.

Auffenberg, Walter. *The Behavioral Ecology of the Komodo Monitor*. Miami: University of Florida–University Presses of Florida, 1981.

Brodkorb, Pierce. "A Giant Flightless Bird from the Pleistocene of Florida," *The Auk* 80, no. 2 (April 1963): 111–115.

Broughton, L. "A Modern Dragon Hunt on Komodo," *National Geographic* l, 70 (1936): 321–331.

Case, Judd A., Michael O. Woodburne, and Dan S. Chaney. "A Gigantic Phororhacoid Bird from Antarctica," *Journal of Paleontology* 61, no. 6 (November 1987): 1280–1284.

Hand, Archer S., and H. Godthelp. *Riversleigh: The Story of Animals in Ancient Rainforests of Inland Australia*. New York: Reed Books, 1994.

Lutz, Dick, and J. Marie Lutz. *Komodo: The Living Dragon*. New edition, 1996. Oregon: Dimi Press. A lecture by Walter Auffenberg is available (in RealAudio) at http://www.si.edu/natzoo/hilights/lectures.htm on the World Wide Web.

Marshall, Larry G. "Land Mammals and the Great American Interchange," *American Scientist* 76, no. 4 (July–August 1988): 380–388.

Marshall, Larry G. "The Terror Bird," *Field Museum of Natural History Bulletin* 49, no. 9 (October 1978): 6–15.

Mourer-Chauviré, Cecile. "Les Oiseaux Fossiles des Phosphorites du Quercy (Éocène Supérieur à Oligocène Supérieur): Implications Paléobiogéographiques," *Geobios, Memoire Special*, no. 6 (1982): 413–426.

Patterson, Bryan, and Jorge L. Kraglievich. "Sistemáticas y Nomenclatura de las Aves Fororracoideas del Plioceno Argentino," *Publicaciones del Museo Municipal de Ciencias Naturales y Tradicional de Mar del Plata* 1, no. 1 (July 15, 1960): 1–51.

Redford, Kent H., and Pamela Shaw. "The Terror Bird Still Screams," *International Wildlife* 19, no. 3 (May /June 1989): 14–16.

Wroe, S. "The Geologically Oldest Dasyurid, from the Miocene of Riversleigh, Northwestern Queensland." In *Paleaontology* (in press). The Riversleigh Society Australian Paleontology site is at http://www.ozemail.com.au/ ~promote1/ auspalaeo/index.html on the World Wide Web.

Wroe, S. "Killer Kangaroo," *Australian Science* 19, no. 6 (July 1998): 25–28.

Chapter 3 | Beasts of the Air and the Sea

Chasing the Ghost Bat

Off to the left of the narrow skiff, past the mangroves at the river's edge, the thick Belizean jungle is an indistinct black mass sliding by under a moonless, starry sky. In the chilly air a faint, salty scent signals that not far ahead lies the mouth of the Sarstoon River, with its treacherous shoals and maze of gill nets. To the right of the boat, the dim yellow lights of fishing huts wink like fireflies.

In the middle of the open boat, a few meters in front of its whining outboard motor, zoologist Bruce W. Miller wields a searchlight. He cuts long, sharp strokes in the gloom over the river. Beside him, his friend and collaborator Michael J. O'Farrell hunches forward on one of the skiff's hard seats. O'Farrell is bone-tired and muddy, and his entire nutritional intake today has been a mug of instant coffee and a few pieces of rye bread.

Suddenly, a large, light-colored bat flits into the bright beam, a ghostly night creature trapped in a sliver of brightness. O'Farrell snaps to attention, and as Miller struggles to keep the light on the dipping, swooping apparition, he and O'Farrell try to make out what species it is. But the animal vanishes, unidentified, within seconds.

O'Farrell sinks in his seat, crestfallen. Not two minutes ago he turned off his Anabat detector, a combination of software and a handheld device that turns an ordinary laptop computer into a sophisticated electronic ear that can record and display the ultrasonic cries of bats. From the shape of these calls, Miller and O'Farrell probably could have determined the species of the mystery bat.

And that, after all, is the main reason they have come to this corner of Central America. Belize is a naturalist's Eden, a nation the size of Massachusetts with fewer than a quarter-million people. But vast tracts of the country, including the entire southernmost Toledo district, have never been surveyed for their biodiversity. Dispersed among Toledo's 4,648 square kilometers (1,795 square miles) of jungle are one highway, a few roads, three small towns and a smattering of villages and camps. Miller and O'Farrell have come to the Sarstoon River to open up Toledo to biological surveying, starting with its bats.

Miller works full-time for the Wildlife Conservation Society, which is paying for most of the expedition. O'Farrell, a freelance biologist who makes his living mostly by surveying government lands for endangered species, paid his own way to be here, as he has four other times in as many years. In that time, Miller and O'Farrell have recorded and identified enough bats in Belize and the Americas to contribute six papers on vocal sequences and population distributions to such publications as the *Journal of Mammalogy*. By their own estimate, they have cataloged the ultrasonic calls of 68 percent of Belize's echo-locating bats.

Earlier in the day Miller and O'Farrell chugged several kilometers up the Sarstoon on the *Meddy Bemps*, an eight-meter-long former lobstering vessel. Miller peered through thick vegetation for little clearings in which they could set up their traps. In most cases, he and O'Farrell catalog a bat's sound—"matching a voice to a face," Miller calls it—by catching the bat, releasing it and recording its emissions as it circles around near the detector.

As the two men scanned the mangroves, Miller explained why the Toledo district is so important. The surrounding jungle, he said, is part of the Selva Maya, the largest block of contiguous tropical forest north of the Amazon. "Belize is a critical link because it's basically undisturbed, and you can have genetic interchange among the critters" through this forest from southeastern Mexico to Panama. Such liberty keeps gene pools from stagnating.

To protect the wildlife from human encroachment, scientists must know which species live here and which ones are truly in trouble. And Belize's 87 known species of bats, Miller explained, account for more than half of the country's mammalian diversity. In addition, he noted, bats play fundamental roles in the forest ecosystem. On a typical night the average insect eater can consume more than 1,000 bugs, including many that would otherwise harm vegetation. Fruit eaters drop seeds as they fly across open areas, stimulating reforestation. And bat guano is manna to the plant kingdom.

But Miller and O'Farrell also admitted that they would hate to leave Toledo without expanding their catalog by matching at least one more voice to a face. After all, they are in an area that has never been surveyed, and it's quite possible that species that are rare up north are common here.

Saccopteryx leptura, the lesser white-lined bat, would be very nice. They are pretty sure it lives in Belize, and if they can confirm its vocal signature, they will have all the species of the genus *Saccopteryx* that are known to live in Belize. And once they nail down *Pteronotus gymnonotus*, the big naked-backed bat, they will have the vocal calls for all the mainland members of the family Mormoopidae.

They could not know then, of course, that their expedition would become a quest for a large white bat, a ghostly, flittering Moby-Dick-in-miniature.

Later that day, after a two-hour search, Miller and O'Farrell found a suitable clearing and set up the traps. The four-meter-tall devices are called harp traps because they snare bats between two sets of vertical monofilament lines.

In the rosy glow just before sunset, the scientists prepared for the night's work. They transferred their gear to the skiff, and motorman Keith Mahler piloted it upriver, keeping near the mangrove-lined bank. O'Farrell, in safari shirt and headlamp, waved the Anabat detector toward the forest. Miller panned the searchlight across a riot of big cohoun palms, provision trees and calliandra beyond the mangroves.

Within minutes, they picked up a bulldog bat, a freetailed bat and a couple of common mustached bats. The Anabat squeaked, chirped and clicked, drawing horizontal lines or scattered dots on the laptop screen as it rendered the bats' ultrasonic squeals for human ears and eyes. To Miller and O'Farrell, intent in the bluish glow of their laptops as their little skiff floated down the pitch-dark river, the squiggles, lines, dots and curves told compelling stories of bats communicating, hunting, feeding and even fishing.

Suddenly, they picked up a short, mysterious signal at around 22 kilohertz, a signal they had seen before but had not yet matched to a "face." "Chances are it's one of the bigger freetails," O'Farrell declared. They detected the call several more times, then checked the traps. In the pouch lay two Mexican funnel-eared bats, a common mustached bat and a Thomas's fruit-eating bat. Back in the skiff, Miller and O'Farrell observed and measured the bats, penciling their findings in a notebook by the light of their headlamps.

Despite the activity, O'Farrell said it had been a quiet night. "We should be picking up long-nosed bats, white-lined bats. We should be getting yellow bats and big brown bats," he complained. But he was intrigued by the 22-kilohertz calls. "We're trying to lay it off on Molossidae," he mused, referring to one of the nine families of bats in Belize. "But we could have one of the big-ass Emballonurids flying overhead."

In half an hour or so he would see a compelling piece of evidence—large and white, flapping over the skiff near the mouth of the Sarstoon. Too bad he would turn off his bat detector minutes before the bat flew overhead.

Mahler zigzags past the gill nets blocking the Sarstoon's mouth and heads toward Amatique Bay. Somewhere out there is the 15-meter ketch *Tempest*, which will be home for the next week if Mahler can find it. As Miller's teeth begin chattering from the damp cold, Mahler homes in on a faint light in the distance. Miller is in luck; it is the *Tempest*.

The next evening they set up the detectors on the ruins of what was once a refreshment stand. Tonight's aerial show features more "feeding buzzes." As an insect-eating bat swoops in on its prey, it increases the rate of its ultrasound pulses until the emissions, as heard through an Anabat, become a short, sharp buzz. The creatures also instinctively shift the pitch of their calls when they are in the company of other bats of the same species, so that they can distinguish their own echoes.

As the evening wears on, the researchers detect a greater white-lined bat, as well as mustached bats, freetailed bats and the mysterious 22-kilohertz signal. Then something different: an unknown member of the family Vespertilionidae, with a signal rising to 58 kilohertz.

It is time to move on to the Temash River, about nine kilometers north. The brief trip along the coast gives the two time to discuss their use of electronic detectors in bat surveying, which is still somewhat controversial. In fact, the only other researcher using the technique in Central America is Elisabeth Kalko of the University of Tübingen in Germany, who uses a custom-built system.

Saccopteryx bilineata
Greater white-lined bat

Harp traps snare bats between two sets of monofilament lines. As they fly toward the lines, bats detect the closer set but can not pick up the set behind. They turn to fly past the first group but collide with the second and fall unharmed into the canvas pouch. The greater white-lined bat (*above*) never wound up in the researchers' traps; they also missed the more elusive lesser white-lined bat.

Anabat, which is commercially available, records and displays only the strongest harmonics of a bat's call. Some critics charge that these dominant harmonics are not enough to distinguish two species of the same genus. Supporters such as Miller retort that all it takes is sufficient practice.

Last year O'Farrell submitted to a three-day blind test conducted by the U.S. Forest Service. Asked to distinguish several species of *Myotis*, he scored 66.7, 76.4 and 84.3 percent correct. A control group—of some graduate students who attempted to identify the same bats using statistical methods—did not score above 68 percent.

"It's almost arrogant to say so," Miller volunteers, "but the technique, as we use it down here, is going to revolutionize the way people sample in the tropics." Traditional bat-surveying techniques, he explains, use only harp traps and mist nets. Neither do well at snagging high flyers, such as Molossidae. Some bat researchers, however, grumble about Miller's and O'Farrell's refusal to release their huge catalogs of Anabat recordings. "It's not that we're unwilling to share," Miller insists. "If [other zoologists] don't have the experience to understand the data, it's not going to help them very much in the beginning. They could misapply the data."

Asked whether sharing the files wouldn't make it easier for more people to get started with the Anabat, O'Farrell argues it would not. Not having the files, he argues, "forces you to go through the same process we go through when we go into a new area—to have to lay hands on the animal and record the calls. It allows you to develop an intuitive feel that's very important."

Michelle Evelyn, a graduate student at Stanford University, says she asked for access to O'Farrell's files and was denied. She later paid $250 to take a two-day Anabat seminar with O'Farrell, biologist William L. Gannon and Chris Corben, the Anabat's inventor. She now defends Miller's and O'Farrell's unwillingness to share their large catalogs of Anabat records. "It's their library, and they spent five years building it. This is totally brand-new stuff, and they're at the forefront. They're publishing their findings; it's not like they're keeping their results secret." Miller and O'Farrell pledge that in time, they will make most of their recordings available, probably through a publicly accessible database now being created at the University of New Mexico.

Motoring 13 kilometers up the Temash, Miller finds only a single spot near the river with enough room to squeeze in the harp traps. A full night's work nets just 29 Anabat files and nothing in the traps. Their chances of extending their catalogs are dwindling. They have only two more nights to try before they must turn the *Meddy Bemps* over to another group of researchers.

The next day they decide to go deep upriver on the Temash, too far to come back to the *Tempest* at night. Sleeping on the hard decks of the *Meddy Bemps* is evidently a more appealing prospect than finishing the expedition without any new voiceprints for their catalogs.

Upriver that night, they decide to split up, each with a bat detector. O'Farrell cruises upriver on the skiff; Miller remains on the *Meddy Bemps*. O'Farrell's Anabat soon starts to chatter as a trio of black mastiffs flutters right over his head in the fading light. The signals come faster and faster until it is clear that O'Farrell is in a "hot spot"—a cloud of insects on which groups of bats are gorging. The data are coming fast and furious: yellow bats, mastiffs, doglike bats and both species of white-lined bat—greater and the elusive lesser—at the same time. Over the next 45 minutes, he also picks up a few signals they haven't yet matched to "faces," squeals that appear to be from a Molossid of some kind and then a member of the family Vespertilionidae. "I have no idea which one," he comments. Then another doglike bat and an unidentified Emballonurid. The creature must be something unknown to Belize, O'Farrell says, because he and Miller thought they had accounted for all the Emballonurids know to live in the country. "We've got to lay hands on these puppies," he adds.

"Whoa!" he yells, after the Anabat crackles yet again. "There's our 22-kilohertz guy, giving us more behavior. That's just what I need."

The frenzy dissipates briefly and then builds to another crescendo as more bats swoop in for supper. Whatever is emitting the 22-kilohertz sig-nal returns for seconds. Still, "our chances of identifying this guy on this trip are slim to none," O'Farrell says. They would have to catch the bat to prove it is the 22-kilohertz caller, and the odds of finding it in the traps are poor.

As the pace quickens, the 22-kilohertz signal sounds through the night several times, discernible only to bats and Anabats. Mahler beams the searchlight into the skies, hoping to spot the animal. Suddenly, he lights up a large white creature, about 15 meters away, flapping toward the palms. "Whoa!" O'Farrell shouts, eyes wide behind wire-framed glasses, veins popping out on his neck. "Look at that! Look at that! He's *white!*" The striking color and the shape of the 22-kilohertz signal, which appears to be that of an Emballonurid, suggest to O'Farrell that the bat is *Diclidurus albus*, the northern ghost bat.

"Nobody has been able to 100 percent identify *Diclidurus*, face and voice," he explains, after he has calmed down a bit. "Elisabeth Kalko in Panama has gotten signals and seen him in a spotlight, like we just did. She's published on hers, but she couched her terms, because she hasn't laid hands on the animal."

At 8:00, on the way back to the *Meddy Bemps*, O'Farrell tallies up the count: in about two hours, he has logged 183 files. Some bats were surely recorded more than once as they circled the area. Nevertheless, a stunning evening.

O'Farrell can hardly contain himself as the skiff approaches Miller on the *Meddy Bemps*. "Oh, boy, Bruce," he begins.

"We got 90-kilohertz Emballonurids," Miller responds.

O'Farrell plays his trump card: "How about a large white bat?"

Miller, absorbed by his laptop screen, seems unmoved. "Mmmm hmmm."

"It's our mystery 20-kilohertz guy," O'Farrell presses on.

"The twenty-two?" Miller says, suddenly very interested.

"That's the guy!" O'Farrell exults.

A moment later Miller is raving about their good fortune. "We've got Emballonurids out the wazoo," he says. "This has been just killer. Everything you could imagine."

The traps have snagged only a single fruit bat. O'Farrell is a little worried. "Elisabeth was saying hers was up around 30?" he asks, meaning kilohertz.

"That's what she told me," Miller responds.

"I'd hate to prove her wrong," O'Farrell says, earnestly.

"Well, there are several species of *Diclidurus*," Miller notes. "But the one collected here [in Belize] was *albus*."

"We have finally hit pay dirt," O'Farrell concludes.

"This is not only the frosting on the cake," Miller agrees. "This is the candles on the cake."

At dawn the next day Miller already has the white bat on his mind. "We have to nail ourselves *Diclidurus*, that's for sure," he says.

Because Miller and O'Farrell know what they want, and because this evening will be their last in Toledo, they decide to forgo the traps and "fish" for the great white bat. They'll bait a hook with a moth and wave it off the end of a long fly rod. With a great deal of luck, they may snare the bat's wing as it tries to sweep the insect into its mouth. The injury to the wing is minor, not unlike the tears the creatures commonly receive as they fly through the bush.

The first 22-kilohertz call arrives at 6:24. Thirteen minutes later there is another. "That's our guy, and he's jumping around," O'Farrell says. "He's starting to vary his calls."

"I'm getting a moth on," replies Miller, up on the bow. "We've got both *Diclidurus* and yellow bat," O'Farrell announces. Seconds later Miller snags a bat on the line, but it's the yellow. He frees the creature, and it flutters away, apparently unharmed. Miller turns for more bait to the huge assortment of moths, flying ants and flies crawling around under the ultraviolet light he's set up on the roof of the boat.

"Biggest-ass moth you can get," O'Farrell suggests.

"I see him, he's up high," Miller reports, as he waves the pole gently, the hooked moth fluttering from the end of its short line.

"He is feeding; he's active," O'Farrell replies, holding the Anabat detector aloft and staring at the laptop screen. "Whoa! He's throwing harmonics up"—aiming a wider ultrasonic signal at the moth to examine it in more detail. "He's foraging! Two good feeding buzzes, Bruce. We've seen just a beautiful amount of this guy's repertoire."

Miller switches bait, to a sphinx moth, an insect about as big as a hummingbird. The moth is so fat Miller can't get the hook all the way through its abdomen, which makes it less likely to snare a bat. Finally, the ghost bat swoops toward the bait but, puzzlingly, does not attack it.

With bugs crawling on his face, Miller looks like a character in a Clive Barker movie. But his attention remains fixed on the white bat, which flutters just out of reach. "I've eaten about half a pound of bugs up here," he gripes. A note of dejection has crept into his voice.

And indeed, before long the big white bat vanishes into the jungle. Miller climbs down from the bow, blinking midges and gnats from under his eyelids. The white bat eluded them this time, but its vocal repertoire is imprinted permanently on their hard disks. That alone warrants several rounds of rum, drunk out of plastic cups until the wee hours.

Ashore at Placencia the next day, they wait for a taxi in the shade of some palms near the dock. "To have laid hands on it would have been a showstopper," O'Farrell says ruefully of the white bat. Nevertheless, both biologists seem fairly certain it was *Diclidurus albus*. "You can't be 100 percent positive," O'Farrell allows. "It may turn out to be a completely different bat, one only known from South America." But they will both probably dream about the great white bat that got away.

They're already planning a trip to Venezuela—it's big, absolutely bat-filled and, like most of Latin America, never explored with an Anabat. Who knows? Maybe they'll even find a few ghosts.

—*Glenn Zorpette, editor,* Scientific American

Running on Water

James W. Glasheen and Thomas A. McMahon

The basilisk lizards of Central America are renowned for their seemingly miraculous flight across water. When startled, these green or brown reptiles scamper over ponds or lakes on their hind legs—the younger ones appearing virtually airborne, the larger ones sinking down somewhat. By videotaping seven *Basiliscus basiliscus* captured in a Costa Rican rain forest and by constructing mechanical models in order to understand the underlying physics, we have been able to decipher the mystery of these lizards' magnificent movements.

It all begins with a slap of the foot. The basilisk lizard strikes the water to create upward force. This force, in turn, provides a medium-size, or 90-gram, lizard with as much as 23 percent of the support it needs to stay on the water surface. Then, a split second later, comes the stroke. As the foot crashes down, it pushes water molecules aside and creates a pothole of air. In addition to the forces generated by accelerating water out of the foot's way, the lizard obtains support from forces created by the difference in pressure between the air cavity above the foot and the hydrostatic pressure below. Together the slap and subsequent stroke can produce 111 percent of the support needed to keep an adult lizard striding along the surface. Smaller lizards, those weighing two grams or less, should be able to create 225 percent of the support they need—and consequently, their runs across the water appear freer and less cumbersome.

All these gains would be lost, however, if the lizard did not pull its foot out of the hole before the water closed in around it. By slanting its long-toed foot backward and by slipping it out while it is surrounded only by air, the creature avoids the drag that would result from pulling its foot through water. A tiny fringe that surrounds the basilisk's five toes may facilitate this motion. Like a parachute, the fringe flares out as the foot is slapped down,

A basilisk lizard sprints across water in the Costa Rican rain forest. Adults usually run on water only when startled; young ones, however, will do so simply to get from one place to another. A medium-size lizard takes about 20 steps a second when running; with each of these steps the lizard's foot creates an air pocket from which the foot is withdrawn before water rushes back in.

thus creating more surface area—all the better to hit the water with. Then, as the foot is pulled up, the fringe collapses, and the long toes are withdrawn just before the hole closes.

As for humans, they have nothing to learn from the lizards except to stay ashore: an 80-kilogram person would have to run 30 meters per second (65 miles an hour) and expend 15 times more sustained muscular energy than a human being has the capacity to expend. The basilisks bask singularly in the liminal world between water and air.

Diving Adaptations of the Weddell Seal

Warren M. Zapol

A person who can swim unaided to a depth of 20 meters and stay submerged for three minutes is considered an expert diver. Yet such an accomplishment pales when compared with that of another mammal, one able to plunge more than 500 meters and remain underwater for more than 70 minutes. This diving virtuoso is the Weddell seal (*Leptonychotes weddelli*), a member of the Phocidae family of true, or earless, seals.

The animal, which flourishes on the shores and coastal ice of Antarctica, plows deep into the cold sea not to set endurance records but in search of food. A quarter of a mile from land, within 50 feet of the 250- to 600-meter-deep sea floor, lives its staple diet: the large Antarctic cod *Dissostichus mawsoni*.

Weddell seals readily withstand water temperatures that fall to -1.9 degrees Celsius by virtue of their large size (adults weigh from 350 to 450 kilograms) and a thick layer of insulating blubber. Diving, which forces the animals to cope with a lack of air and with intense undersea pressure, constitutes a more complex challenge. Indeed, unraveling the adaptations to this challenge has required decades of laboratory research by many investigators and, more recently, a spate of field studies. The field studies suggest that certain long-held beliefs based on laboratory studies may need to be modified. Forcing a seal confined in a laboratory to put its face underwater does not necessarily evoke the same response as a dive undertaken freely in the sea.

The specific problems posed by diving are considerable. Above all, the seal must provide its tissues with oxygen. At the same time it must limit the buildup in the blood of carbon dioxide, a by-product of the oxidation of glucose for energy. This gas is generated by the tissues and then carried in the blood to the lungs for removal. When an animal is submerged, the gas can accumulate in the blood, upsetting the fluid's delicate pH balance.

The animal also has to avoid the many ills that extreme pressure can cause. For every 10 meters of depth, an animal or a person is subject to an additional "atmosphere" of external pressure—that is, to the push exerted at sea level by a 14.7-pound force on one square inch of area, or the pressure exerted by a 760-millimeter column of mercury. One potential effect of underwater pressure is an increase in the excitability of nerve cells, which can result in convulsions. Pressure also squeezes air pockets, such as the air sinuses in the human head. The squeezing can cause pain, and if the body cannot deliver enough air to the pockets to equalize the external pressure, blood vessels may expand into the air spaces and burst.

Pressure also compresses gases, posing a danger when it affects the nitrogen in the alveoli, the tiny gas sacs in the lungs. (Body fluids and fluid-filled organs are compressed only minimally underwater.) Nitrogen gas constitutes some 78 percent of the air. Normally it passes harmlessly into the circulation, but when the air in the lungs is put under great pressure, as it is during descent, excess nitrogen dissolves in the blood and tissues; it may then lead to narcosis, a disorder that divers call rapture of the deep. Narcosis is identified by such symptoms as intoxication, loss of coordination and vision, drowsiness and unconsciousness. During ascent, a too rapid trip to the surface can cause the nitrogen tension in the blood and tissues to be greater than the external pressure on the body. Then dissolved nitrogen may come out of solution and bubble ("the bends"). In addition to producing pain in the joints and elsewhere, the bubbles may block vessels in the brain and spinal cord, leading to paralysis and even death.

Laboratory studies, in spite of their limitations, have revealed many of the strategies by which the diving seal appears to ensure an adequate supply of oxygen and avoid the disorders described above.

One major and still undisputed laboratory finding is that the seal stores an abundance of oxygen—almost twice as much per kilogram of body weight as a human being does. It also concentrates the oxygen where it is most needed during a dive: in the blood and, to a lesser extent, in the muscles. People are particularly dependent on the lungs for oxygen, keeping 36 percent of their total supply in the lungs and 51 percent in the blood, but the seal stows only 5 percent in the lungs and a full 70 percent in the blood.

Similarly, a person stores just 13 percent of its oxygen in the muscles, but the Weddell seal keeps about 25 percent there, bound to the oxygen-carrying pigment myoglobin.

Vast amounts of oxygen can be maintained in the seal's blood in part because the volume is enormous. In 1969 Claude J. M. Lenfant discovered that in contrast to the blood of human beings, which typically accounts for 7 percent of the body weight, the blood of the Weddell seal accounts for 14 percent of the animal's weight. (This comparison actually underestimates the amount of blood that is available to working tissues, because blubber, which constitutes about a third of the animal's mass, receives little blood.) Moreover, the seal's blood has great quantities of hemoglobin, the oxygen-carrying pigment of red blood cells. When my group drew blood from seals in the laboratory, we found that red cells accounted for some 60 percent of the volume of each drop; in human beings these cells occupy only from 35 to 45 percent of the volume.

Although the Weddell seal's oxygen supply is impressive, it is not infinite. Like other diving animals, the seal has therefore devised ways to conserve its fuel. When any mammal puts its face in the water, neural impulses trigger the brain to induce the so-called diving reflex: as the animal stops breathing, bradycardia (a slowing of the heart rate) ensues and certain arteries become constricted, limiting the blood that flows to the organs they feed.

The rapid onset of bradycardia at the start of a dive has been recognized in animals for more than 100 years. It happens in human beings but appears to be most profound in species that dive habitually, such as seals and whales. A slowed heart rate is beneficial underwater because it enables the heart to work less hard and hence to require less oxygen. Bradycardia also reduces the heart's output of blood, a change that helps to keep blood pressure at a normal level when the arteries are constricted. Furthermore, as the flow of blood diminishes, the metabolism slows, reducing the oxygen needs of tissues throughout the body.

Constriction of the arteries presumably ensures that the maximum supply of blood, and therefore of oxygen, will be available to the tissues that are most crucial to survival. My colleagues and I measured blood flow to various tissues in the seal during a laboratory dive. Consistent with earlier findings, we found that the seal continued to supply blood at a normal rate to the retina, brain and spinal cord, all of which are vital to navigation and motor control. (As would be expected, the heart received blood but the amount was reduced to match the organ's reduced workload.)

Two other tissues also continued to benefit from a normal flow: the adrenal glands and, in pregnant seals, the placenta. Just why the body perfuses the adrenals is not clear, but the fact that the glands produce high levels of the hormone cortisol may provide a clue. Some evidence suggests that cortisol serves to stabilize nerve cells during a dive, thereby preventing pressure-induced convulsions. Why blood flows to the placenta is more obvious. This organ is vital to fetal gas exchange and must continue to function if the submarine within a submarine is to survive.

Our studies of oxygen distribution also confirmed that the seal essentially shuts off the flow of blood to most other organ systems and tissues during laboratory dives. When this flow ceases, many of the affected tissues (such as the kidneys) stop functioning until the animal comes up for air. Certain other tissues apparently switch to anaerobic, or oxygen-independent, metabolism if they have crucial tasks to perform. The telling by-product of anaerobic metabolism is lactic acid; when seals surface after forced dives, the levels of lactic acid in the blood soar above the resting state.

Although initiating anaerobic metabolism can be important when oxygen is lacking, it can also be extremely dangerous. High levels of lactic acid lower the pH of the blood and can lead to acidosis, which may cause cramping, a weakening of the heart's ability to contract and even death. Laboratory studies by P. F. Scholander, who studied the diving reflex in the 1930's, suggested that the seal avoids acidosis by confining anaerobic metabolism to the skeletal muscles and other tissues that are isolated from the blood supply during laboratory dives. With the blood flow shut off, these tissues cannot release lactic acid into the blood until the animal surfaces. At that time the liver, lungs and other organs can clear out the by-product.

Laboratory work has also attempted to explain how the Weddell seal handles external pressure. In addition to raising the possibility that elevated cortisol levels may prevent convulsions, the studies have shown that the seal lacks the potentially troublesome air sinuses of other mammals. The seal likewise has ways of avoiding nitrogen narcosis and the bends. Its lungs are small for its weight, and so they have a reduced capacity for storing nitrogen that might diffuse into the blood in the course of a dive. Moreover, the animal exhales before submerging. The obvious effect is to reduce the buoyancy that impedes descent, but exhalation has the added benefit of reducing gas volume in the lungs still further.

During a dive, seawater pressure on the animal's collapsible rib cage undoubtedly squeezes most of the remaining nitrogen out of the alveoli and into the bronchial air-duct system. According to anatomical studies done by Gerald L. Kooyman and his co-workers, the seal's bronchi and bronchioles are supported by rings of cartilage that enable the airways to serve as an armored gas-storage reservoir. Because these passages, unlike the alveoli, have no direct contact with the blood, they do not introduce nitrogen into the circulation. (Some oxygen is certainly sequestered in the seal's airways as well, but not much; only 21 percent of inhaled air is oxygen.) In contrast, the bronchi and bronchioles of human beings would close down under intense pressure and so could not store excess nitrogen.

Kooyman also determined, on the basis of forced dives in a compression chamber, that the seal's lungs collapse when the animal reaches a depth ranging between 50 and 70 meters. Collapsed lungs would halt the flow of nitrogen into the blood and hence limit the total amount of nitrogen that accumulates there.

Kooyman and his co-workers carried out some field studies that caused many of us to wonder whether the seal in the ocean responds to a dive the way it does in the laboratory. When Kooyman's group attempted to study seals engaged in voluntary dives in the ocean, their data suggested that the animals may not always exhibit a pure diving reflex. However the available equipment could not monitor complete dives.

After Roger D. Hill developed software and constructed a battery-operated, eight-bit computer that would make it possible to evaluate the seal's physiological and metabolic responses throughout free dives at sea, a group of us from laboratories around the world converged on the research station on the shore of McMurdo Sound in Antarctica.

Hill's diving computer, which was encapsulated to withstand 500-meter depths, did everything but steer. It recorded heart rate and depth at predetermined intervals for several days. It also controlled an electric pump that took up to seven arterial-blood samples at specified times (such as 10 minutes into the dive) and depths. After collecting the samples the computer pumped the blood into a bag or syringes tethered to a six-foot fiber-optic line, which itself had additional functions when the seal surfaced.

We gathered seals from nearby colonies and sledged them to the study site, a hole three feet in diameter drilled through ice six feet thick. There we anesthetized a subject with harmless techniques devised by Robert C. Schneider, inserted the necessary catheters and attached the computer, which was fastened to a rubber sheet glued to the seal's dorsal fur. (When the animals molted later in the summer, they readily shed the rubber appendage.)

Once rigged with our computer and recovered from general anesthesia, the seal was free to enter the hole and swim off. We were confident it would return with blood samples and data because we adopted a rather reliable tactic devised by Kooyman. Knowing that Weddell seals can swim only a few kilometers underwater, he studied the animals at isolated holes drilled in broad ice sheets; the seals had to return to the site of origin in order to breathe.

When the seal returned, we quickly connected its fiber-optic line to the stationary computer. Within 10 seconds the larger machine collected data stored in the diving computer and, when appropriate, gave it new instructions.

We found, as Kooyman had earlier, that some 95 percent of the seal's voluntary dives last for less than 20 minutes. These tend to be feeding dives in which the animals head directly for their prey and then return. The seals embark on the 5 percent of dives that last longer than 20 or 30 minutes when they explore distant routes or must escape from predators.

Studying the all-important distribution of oxygen to tissues, Peter W. Hochachka showed that the seals do not release lactic acid into the circulation during or after sea journeys that last up to 20 minutes. This indicated that during short natural dives—that is, the majority of the seals' journeys—the muscles do not resort to the anaerobic metabolism observed in laboratory dives and must receive some blood. (The muscles would probably account for most of the lactic acid in the blood because they are abundant and also do work when the animal dives.) With little or no lactic acid to break down after a dive, the seal often resumes fishing within minutes after taking a few breaths at the surface.

In contrast to the feeding dives, the seal's occasional long excursions do evoke the classic diving response seen in the laboratory. The long forays are characterized by profound bradycardia with little variability of heart rate. After (but not during) such dives the Weddell seal releases lactic acid into its blood, indicating that the animal shuts off the blood flow to its muscles and meticulously conserves oxygen while diving. Having switched to anaerobic metabolism in the muscles, the seal can stay underwater for an hour or more. It pays a price, though: when it finally surfaces, it does not dive again until it has cleared away the lactic acid released by the muscles, a process that can take up to an hour. Why do even short laboratory dives elicit a response characteristic of long field dives? In the laboratory the seal does not know how long it will be submerged and so prepares for the worst.

Other studies of oxygen delivery indicate that early in both feeding and exploratory dives at sea the seal actually increases the concentration of red blood cells in the circulation, thereby maximizing the hemoglobin level in the blood and thus the amount of oxygen available to the tissues. In a remarkable discovery Jesper Qvist, who was a part of our Antarctic team, found that the red-cell concentration in the arteries increases by 50 percent in the first 10 to 15 minutes of a dive. In contrast to my group's laboratory findings, which suggested that the levels are always high, Qvist showed that the cells initially account for only 35 or 40 percent of the circulating blood volume and then rise to 60 percent during a dive. (The levels return to normal within 10 minutes after the animal surfaces.)

Where might the bounty of new cells come from? The spleen is a reasonable guess. This poorly understood organ is known to contract when the sympathetic nervous system is activated, as it is when a mammal dives or is frightened. (Indeed, fear may explain why some seals have elevated red-cell levels when they are confined in a laboratory.) Contraction of the spleen could well inject stored oxygen-rich red cells into the seal's highly expand-able venous system; the heart could then deliver them to the arterial circula-tion as needed. After the seal returned to the surface for air, the circulating red cells would be readily reloaded with oxygen and stored again.

An oxygen-supplying role for the spleen is not unprecedented; the organ is known to infuse red blood cells into the circulation within minutes after a racehorse begins intensive exercise. On the basis of organ size and the degree to which hemoglobin levels rise during a dive, my colleagues and I estimate that the Weddell seal warehouses approximately 60 percent of its total red-cell supply in the spleen, whereas a human maintains less than 10 percent there. Indeed, the seal's spleen appears to be something of a contrac-tile scuba tank in its ability to store and release oxygen needed for a dive.

The effects of an oxygenated red-cell infusion become particularly apparent in long dives. In the period when the red-cell concentration is ris-ing, the blood's oxygen content remains constant, indicating that the amount consumed by the brain, heart and other crucial tissues is somehow being replaced. The same plateau is not seen in feeding dives, where the muscles consume oxygen; then oxygen levels fall steadily. In this instance the oxygen burned by the seal's muscles probably outstrips the ability of the splenic blood-storage system to inject red cells into the circulation.

In addition to providing oxygen during a dive, the inflow of fresh red blood cells into the circulation probably serves another important purpose: the dilution of gases dissolved in the blood. Such an effect would explain why the carbon dioxide concentration rises surprisingly little in the course of field diving. Dilution would also help to explain why nitrogen does not cause narcosis or the bends in the seal. The field studies suggest other expla-nations as well; some nitrogen diffuses out of the blood and into muscles and blubber.

As a whole our field studies demonstrate that the Weddell seal's responses to its occasional long dives look much as laboratory dives predict they should. The diving reflex is in full force: the heart rate slows and remains low throughout the dive, and the muscles switch from aerobic to anaerobic metabolism, indicating that their supply of blood is shut off, probably because their arteries are constricted.

In the majority of dives, in contrast, the profile is rather different. When the seal embarks on a feeding excursion, the diving response is modified. The heart rate slows but is more variable, speeding up as the seal swims faster. Moreover, the muscles continue to rely on aerobic metabolism; apparently they continue to receive some blood, indicating that vascular constriction is modulated. Early in its dive the seal apparently "decides" whether its foray will be long or short and whether or not it must resort to draconian measures to conserve oxygen.

Stalking the Wild Dugong

Madhusree Muskerjee

From the porch where I am slumped, exhausted by the heat, I stare in astonishment at a man walking up the forest trail from the beach, snorkel dangling from one hand. I have just arrived at Dugong Creek, a remote corner of Little Andaman Island in the Bay of Bengal, to meet the Onges, a group of hunter-gatherers believed to be descended from Asia's first humans. I hadn't expected to find other visitors.

"You know there are crocodiles," I say, indicating his snorkel.

"A hazard of the trade," he grins.

Himansu S. Das is a sea-grass ecologist. Because dugongs, Old World relatives of the manatee, feed on underwater greenery, he had guessed that Dugong Creek would have beds of sea grass nearby. The animals themselves, though, were likely to be long gone. Once seen in the hundreds or even thousands along the tropical coasts of Africa and Asia, these sea elephants are all but extinct in most of their range and occur in reasonable numbers only in Australia. In five years of exploration, Das has gathered evidence of at most 40 dugongs throughout the Andaman and Nicobar archipelago. To his surprise, he has just learned from the Onges that a family of four still lives in Dugong Creek, down one since their hunt of two weeks ago.

The grass beds nourish not only these rare mammals but also marine turtles and a variety of fish and shellfish. With the help of a grant from UNESCO, Das is estimating the impact of humans on the ecology. In fact, it is the local peoples who point him to the beds, more predictably than do the satellite images on which he initially relied.

The next afternoon, under a blistering sun, we set out for an Onge camp a kilometer or so along the shore.

The Onges, we discover, have not seen a dugong but have harpooned two turtles. One is being cooked, and the other is secured at the end of a long rope stretched into the sea. When Koira, an Onge man, pulls on the leash, a head sticks anxiously out of the water as the animal looks to see where it is being drawn. It is an endangered green sea turtle, small, about 15 kilograms.

Neither of us begrudges the Onges their meal. They have lived on Little Andaman for millennia with no harm to its biodiversity and now, because of pressure from recent settlers, will probably vanish long before the turtles. The main threat to the sea-grass beds and to the creatures that depend on them is the silt that muddies the water as the dense tropical forest is cut down: the marine plants die of darkness. Overexploitation of fish, shellfish and other marine species by immigrants from mainland India and by fishers from as far away as Thailand is another pressing problem.

As the grass patches shrink, the dugongs become confined to ever smaller regions that are also the local fishing grounds. Some fishers set their nets around the beds to catch predators, such as sharks, that come to feed on smaller fish, but the nets entangle turtles as well as an occasional dugong. Das will be recommending to the Indian authorities that some sea-grass beds be protected as sanctuaries. But as we return I realize with sadness that it's already too late for the Andaman dugong.

Secrets of the Slime Hag

Frederic H. Martini

Thump. After an hour of descending through near-total darkness in the research submarine *Alvin*, we slide into the silty ocean bottom roughly 1,700 meters (just over one mile) below the surface of the Pacific, off the coast of Southern California. The pilot switches on the floodlights, illuminating a dense cloud of sediment kicked up by our arrival. Several minutes pass as we wait for the debris to settle and activate the sub's sonar system, which shows a large target roughly 240 meters away. As we move closer, we see through *Alvin*'s portholes the ghostly white carcass of a 32-metric-ton gray whale. The whale's watery grave is anything but peaceful: it is swarming with hundreds of half-meter-long hagfishes, which are methodically gnawing away at the whale's chalky blubber, bite by bite.

Scenes like this are eerie enough to keep some people up at night—and they change forever one's concept of burial at sea. But for my colleagues and me, who study the biology of hagfishes, they provide a fascinating glimpse into the lives of these strange and slimy animals. For years, the habits of hagfishes—which are sometimes called slime hags—and their place on the evolutionary tree of life have been a matter of conjecture. But hagfishes in many ways resemble the first craniates (animals with a braincase). The evolutionary path leading toward humans—and all other vertebrates (animals with a backbone)—probably diverged from that of hagfishes 530 million years ago. Also, hagfishes are much more abundant—and probably play a much more important role in the ecology of the ocean-bottom community—than anyone would have guessed a decade ago.

Slime Balls of the Sea

The word "slimy" can only begin to describe the average hagfish: one good-size adult can secrete enough slime from its roughly 200 slime glands to turn a seven-liter bucket of water into a gelatinous mess within minutes. Hagfishes release slime in varying amounts, depending on the circumstances. They tend to produce slime in small amounts while feeding on a carcass, a behavior that might be designed to ward off other scavengers. But when attacked or seized, a hagfish can ooze gobs of goo, from all its slime glands at once. Although the slime is initially secreted as a small quantity of viscous, white fluid, it expands several hundred times as it absorbs seawater to form a slime ball that can coat the gills of predatory fish and either suffocate them or distress them enough to make them swim away. But for all its utility, the slime appears to be equally distressing to the hagfish. To rid its body of the sticky mucus, a hagfish literally ties its tail in a knot and sweeps the knot toward its head to scrape itself clean.

People often mistake the hagfish for an eel because both animals are long and cylindrical. The common names of several species of hagfish even include the term "eel," usually accompanied by a descriptive adjective ("slime eel," for example). As is so often the case, however, such common names are misleading. Hagfishes are not eels at all: true eels are bony fish with the requisite prominent eyes, paired pectoral and pelvic fins, a hard skeleton, bony scales and strong jaws. Like other bony fish, eels rely for respiration on gills that are attached to bones called gill arches and covered by a bony flap called an operculum.

In contrast, hagfishes are much simpler in form and function. They lack true eyes and paired fins, and their rudimentary skeleton consists only of a longitudinal stiffening rod made of cartilage, called the notochord, and several smaller cartilaginous elements, including a rudimentary braincase, or cranium. Hagfishes do not have scales; instead they have a thick, slippery skin and large, complex slime glands. In addition, they lack jaws, and their gills are a series of pouches that are different from the gills of any other living fish.

Hagfishes can be found in marine waters throughout the world, with the apparent exception of the Arctic and Antarctic seas. Although the animals always live near the ocean bottom, they can survive at a variety of depths. Water temperature is the primary factor that limits the habitat of hagfishes: they appear to prefer waters cooler than 22 degrees Celsius (71 degrees

Fahrenheit). In the cold coastal waters off South Africa, Chile and New Zealand, the animals sometimes enter the intertidal zone, where they have been collected in tide pools as shallow as five meters. In tropical seas, though, hagfishes are seldom seen at depths shallower than 600 meters.

There are roughly 60 species of hagfishes, most of which are members of two major genera: *Eptatretus* or *Myxine*. (Many species in these genera, however, are known only from single specimens.) The genus *Eptatretus*, with roughly 37 species, includes the largest hagfish known, *E. carlhubbsi*, which can reach a length of 1.4 meters and can weigh several kilograms. Underneath their skin, *Eptatretus* species have the evolutionary remnants of eyes that are covered by translucent eyespots. Their heads also bear traces of lateral lines, sensory structures that extend down the sides of bony fish. Individual *Eptatretus* live in long-term burrows in the ocean floor but may roam widely among rocks or other hard substrates.

Members of the genus *Myxine*, which includes roughly 18 species, are more specialized than *Eptatretus* for living in burrows. *Myxine* are generally more slender, have even more degenerate eyes that lack eyespots, and show no traces of lateral lines. Typical *Myxine* live in transitory burrows and are always found in or near soft, muddy sediments.

The feeding habits of hagfishes—which can eat small, live prey and act as scavengers—are particularly distinctive. As a hagfish feeds, it protrudes a very effective feeding apparatus consisting of two dental plates, each supporting two curved rows of sharp, horny cusps. The dental plates are hinged along the midline, allowing them to open and close like a book. To take a bite, a hagfish extends its feeding apparatus, causing the "book" to open, and presses the dental plates against a fleshy surface—whether it be the body of a sea worm, a dead fish or your hand. When the hagfish withdraws the apparatus, the book closes and the opposing cusps grasp and tear the flesh, carrying it into the mouth. (The fang situated above the dental plates keeps live prey from wriggling away between bites.)

This feeding method works quite well when a hagfish preys on thin-skinned, soft-bodied sea worms, but the cusps cannot pierce the scales of fish or the skin of whales. Unless other scavengers have already opened the way, when feeding on a large carcass a hagfish usually takes the easy route, entering the body through the mouth, gills or anus. It then consumes the soft tissues from within, until only the bones and skin remain. More than one disappointed fisher has hauled in a prize fish that turned out to be a hollow shell full of hagfishes.

Only a few details are known about hagfish reproduction. Hagfish gonads form in a fold of tissue on the right side of the abdominal cavity. In a female an ovary forms in the anterior two thirds of the fold; in a male, a testis forms in the posterior third. Curiously, individuals with both types of gonads are found occasionally. Females, which in some species outnumber males more than 100 to one, produce between 20 and 30 yolky, shelled eggs at a time. There are no oviducts; mature eggs are released into the abdominal cavity. The eggs—which vary in size from 20 to 70 millimeters, depending on the species—usually have hooked filaments at either end that enable them to lock together and be ejected in a chain. In males the testis produces sperm in follicles that release sperm into the abdominal cavity. Eggs or sperm then leave the abdominal cavity through a large pore into the cloaca, an excretory chamber that also receives and expels urinary and digestive wastes.

Living Fossils

Given the bizarre and mysterious biology of hagfishes, it is no wonder that these blind, jawless, scaleless, finless, bottom-dwelling creatures were not immediately recognized (or acknowledged) as distant cousins of humans.

Today, scientists recognize that hagfishes are virtual biological time machines. A fossilized hagfish, *Myxinikela*, was found in sediments deposited roughly 330 million years ago. Aside from *Myxinikela*'s large eyes, if it were alive today it could easily pass for a modern hagfish.

Biologists neglected hagfishes for much of the past century primarily because of the way they classified animals. Until relatively recently, they relied on common features, such as the presence or absence of eyes or jaws, to establish relatedness between creatures. Under this scheme, hagfishes were lumped together with lampreys in a group called either Agnatha (literally, "no jaws") or Cyclostomata ("round mouths"). Hagfishes and lampreys were classified together because both lack jaws, paired fins, a bony skeleton and scales. Because the habitats of hagfishes make them relatively hard to come by, biologists concentrated on lampreys, which spend part of their lives in freshwater streams and rivers and therefore are much easier to catch.

In more recent years, the acceptance of phylogenetic systematics, or cladistics—classifying animals according to shared, specialized characteristics—has forced a reevaluation of the old methods for deciding what is related to what. Biologists now recognize that it is impossible to tell whether the ancestors of a given organism never had a particular feature or whether they had the feature but their descendants simply lost it sometime during evolution.

According to cladistics, hagfishes and lampreys are separate and distinct groups within the chordates (Chordata). At some point in their lives, all chordates display the following characteristics: a hollow, dorsal nerve cord; a notochord, situated immediately below the nerve cord; gill slits; and a segmentally muscled tail that extends past the anus. Hagfishes are considered the most primitive living craniates. Lampreys also have a cranium, but unlike hagfishes, they also have segments of cartilage to protect their nerve cord. These cartilage segments are the first evolutionary rudiments of a backbone, or vertebral column. Lampreys, therefore, are considered the most primitive living vertebrates.

By comparing fossils with living creatures, biologists can create diagrams called cladograms that display the evolutionary relations among organisms. A cladogram of the chordates suggests that the hagfish diverged

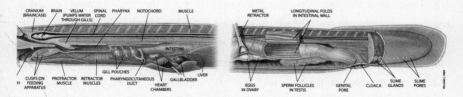

Anatomical views of the anterior (left) and posterior (right) parts of a Pacific hagfish highlight both the animal's unique specializations and other, more general characteristics—such as a cranium—that persist in more evolved animals. (The middle—roughly one-third of the animal's length—has been omitted.) Like a small proportion of most species of hagfishes, this specimen has both an ovary and a testis.

from the vertebrate evolutionary line around 530 million years ago. It also reveals that the predecessors of the hagfish never had a bony skeleton but that those of the lamprey did. What is more, the cladogram suggests that all early craniates had a complex protrusible feeding apparatus comparable to that of hagfishes. Early vertebrates, including the distant ancestors of humans, probably shared many other anatomical and physiological characteristics with modern hagfishes. But hagfishes have evolved many unique specializations: their eyes and lateral lines regressed, and they developed slime glands.

Besides their key position on the tree of life, hagfishes are also gaining new respect as members of the complex ecosystem of the ocean bottom. Scientists now know that the animals are more abundant than was once thought. Based on trapping surveys done between 1987 and 1992, my colleagues and I estimated that the inner Gulf of Maine contains population densities of up to 500,000 *M. glutinosa* per square kilometer.

We also now recognize the degree of hagfish predation on the populations of other animals that live near the bottom of the ocean. Although individual hagfishes have extremely low metabolic rates, their energy needs add up. The average number of *M. glutinosa* that inhabit one square kilometer of seafloor (59,700 animals) must consume the caloric equivalent of 18.25 metric tons of shrimp, 11.7 metric tons of sea worms or 9.9 metric tons of fish every year. And that amount would be sufficient only to keep the animals alive at rest; when they swim or burrow, their energy demands increase between four and five times.

Hagfishes also consume discarded so-called bycatch from commercial trawl-fishing fleets and play a central role in recycling the carcasses of dead marine vertebrates, including whales. Craig Smith of the University of Hawaii has found that hagfishes can remove roughly 90 percent of the energy content of small packages of bait sunk to the seafloor at depths of 1,200 meters. But hagfishes are not just important ecologically for their roles as predators or scavengers; they also serve as important prey for a surprising number of marine animals.

The Slime Hag Trade

In many locales around the world, hagfishes have become the focus of a large and flourishing commercial fishery. Since the 1960s there has been a booming trade in leather goods produced from tanned hagfish skin.

The demand for suitable skins has supported commercial hagfish ventures around the Pacific Rim and in the western North Atlantic. The collection method is very low tech: multiple traps baited with anything from herring to kitchen scraps are set along a line on the sea bottom and left overnight. The traps can be 19-liter pails with lids or 190-liter barrels with small holes in the sides; once inside, most hagfishes become trapped in the bait and their own slime. In previously unfished areas, more than 100 hagfishes have been found to enter a given trap during its first hour on the bottom.

Unfortunately, the demand for hagfish skin has depleted the populations of many species because the trapping rate has far exceeded their rate of reproduction.

In the region of New England, annual landings of hagfishes went from virtually zero in 1991 to roughly 1,950 metric tons in 1996. Over that five-year span, roughly 50 million hagfishes were processed and shipped overseas. By 1996 there were signs that hagfish fisheries were in trouble; recent declines in landings, average size and catch per trap suggest that the trouble is serious.

Although hagfishes are more plentiful than once thought, we do not yet know enough about them to manage a sustainable hagfish fishery. In the meantime, we should take simple steps—such as requiring holes in commercial traps through which small, young hagfishes could escape—to reduce the impact on hagfish populations. When we drastically reduce the number of any species—even the lowly (and to some, loathsome) hagfish—we are performing an ecosystem experiment on a grand scale. As usual, we cannot yet begin to predict the eventual results.

The Giant Squid

Clyde F. E. Roper and Kenneth J. Boss

In *Moby Dick* Herman Melville describes a sea creature of "vast pulpy mass, furlongs in length…, long arms radiating from its center and curling and twisting like a nest of anacondas." He apparently had in mind the giant squid; his description reflects the meager information that was available about the animal in his time. Indeed, until the crew of a French warship sighted one in 1861 and managed to haul in part of it the animal was quasi-mythical. Even now, when a considerable number of specimens have been reported, the giant squid (weighing up to 1,000 pounds and having an overall length of some 18 meters, or nearly 60 feet, if the tentacles are extended) remains largely mysterious. No living specimen has ever been maintained in a research institution or an aquarium. Most of what is known comes from strandings, when the squid is dead or dying; from capture by fishermen of animals that soon died; and from specimens removed from the stomach of toothed (fish- or squid-eating) whales. On the basis of this information one can state what a mature giant squid looks like and can say quite a bit about its internal anatomy.

Teuthologists, the specialists who study cephalopods (the group of marine animals that includes the squid, the cuttlefish and the octopus), have placed the giant squid by itself in the genus *Architeuthis* of the family Architeuthidae. Although authorities differ, it has been suggested that the 19 nominal species can in fact be encompassed by only three: *Architeuthis sanctipauli* in the southern hemisphere, *A. japonica* in the northern Pacific and *A. dux* in the northern Atlantic. Some evidence, based mainly on the relative size of the head and the outline of the caudal fin, has been marshaled to indicate that as many as five species live in the Atlantic, two in the northern Pacific and several more in the southern Pacific.

In 1980 a specimen of medium size—about 10 meters in total length—

was stranded on Plum Island in Massachusetts. The animal was preserved and studied in detail. The study provides much of the data for the following description, in which we adhere to the standard method for avoiding ambiguity about anatomical positions by relying on the terms dorsal for back or top, ventral for front or bottom, caudal for at or near the tail, anterior for forward, posterior for rearward, proximal for near a reference point and distal for far. The normal swimming position of the giant squid is horizontal, so that the animal's dorsal side is at the top and its ventral side is at the bottom.

The giant squid's cylindrical head may reach nearly a meter in length; it is connected to the body proper by a "neck" bearing a circumferential collar or sheath with a dorsal locking cartilage. A crown of eight thick, muscular arms and two very long and thin but muscular tentacles surrounds the buccal (mouth) apparatus and extends from the anterior end of the head. (Cephalopod means head-foot.)

Although the arms are proportionately long in young individuals, they are relatively much shorter than the tentacles in adults. An arm may attain a circumference of 50 centimeters at the base and a length of three meters. Each arm bears on its inner surface low, weak and sometimes scalloped protective membranes that border two rows of suckers. The suckers gradually decrease in size distally until at the tip of the arm they are merely tiny knobs.

The two ventral arms in males are hectocotylized; that is, specialized to facilitate the fertilization of the female's eggs. With them the male transfers spermatophores to the female in mating. These arms differ among species in their length and diameter and in the extent of their modification for the mating function. Distally the arms bear, as continuations of the two rows of suckers, two rows of rectangular pads separated by a deep furrow.

The two tentacles, with which the squid seizes its prey in a motion rather like the thrusting and closing of a pair of pliers, are about 25 centimeters in circumference at the base and can reach a length of more than 10 meters. The stalk of a tentacle is bare along the base; alternating small suckers and adhesive knobs may appear farther along its length. The suckers and knobs increase in size and frequency toward the club: the slightly

Adult giant squid can be as much as 18 meters (60 feet) long and can weigh as much as 450 kilograms (1,000 pounds). This drawing is based on numerous specimens or parts of them that have been drawn, photographed or preserved over the past century.

expanded distal end of the tentacle. The manus, or palm, of the club has four rows of finely toothed suckers. The larger medial ones (in two rows) are about two and a half times the diameter of the smaller ones in the marginal rows. The diameter of a tentacular sucker may reach 5.2 centimeters. The distal end of the club, the dactylus (finger), is pointed and attenuated but is covered with hundreds of small suckers.

All the suckers of a giant squid are shaped like a suction cup. Each sucker is set on a muscular pedicle, or short stalk, that can be moved by the animal. The perimeter of a sucker is rimmed by a sharply toothed ring of chitin (the hard material that forms the outer covering of many crustaceans and insects) that adheres to the surface of the prey when the sucker is applied. No specialized hooks appear in place of suckers, as they do in certain other oceanic squids.

Imprints or scars from squid suckers have been found on the skin of sperm whales and even in their stomachs. The diameter of these scars has been variously exaggerated, sometimes to as much as 20 centimeters (eight inches). Such reports have led by extrapolation to a distorted estimate of the maximum length of giant squids. The most reliable evidence suggests that the average diameter of the suckers on the arms is about 2.5 centimeters and that the maximum diameter is 5.2 centimeters, the size sometimes found on a tentacular club.

The mantle, or body, of the giant squid is more or less narrowly cone-shaped. It tapers to a bluntly pointed tail. In adults a short, stout, taillike

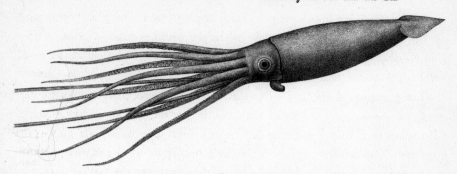

In the class of cephalopods the giant squid is placed in the genus Architeuthis; the number of species in the genus has yet to be firmly established.

projection extends beyond the fins; juveniles do not have it. The fins are flexible but not strongly muscular, suggesting that they serve as stabilizing vanes.

As a modification of the molluscan foot a large, muscular funnel rises ventrally behind the head at the anterior end of the mantle. With it the squid propels itself by squirting water out of its mantle cavity. The funnel is highly mobile, so that the squid can dart forward, backward, up, down or to the side. Inside the funnel is a flaplike valve that prevents the backflow of water between squirts. Ventrally at the base of the funnel on each side is a

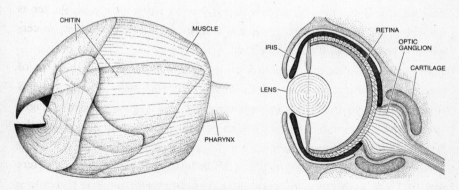

The mouth and eye of the giant squid are among the animal's noteworthy features. The powerful beak consists of chitin, the hard material that forms the external covering of many crustaceans and insects. With its beak the squid cuts its prey into pieces it can swallow. The eyes of the squid are the largest in the animal kingdom, each one being approximately 25 centimeters (10 inches) in diameter.

groove of cartilage that interlocks with corresponding cartilaginous ridges on the inner surface of the mantle when the squid is expelling water. This funnel-locking mechanism prevents water from escaping around the neck, forcing all the water to go through the funnel.

The giant squid also has a translucent internal supportive structure variously called the gladius or the pen. It is the remnant of an archaic internal calcareous shell that is still found in more primitive cephalopods such as the cuttlefish. The gladius lies in a sac in the musculature of the mantle, extending posteriorly from the anterior edge of the mantle to the taillike posterior extremity. It functions as a skeletal rod for the attachment of muscles and as a supporting staff for the elongated body.

The giant squid's eyes are enormous, larger than the headlights of an automobile. With a diameter approaching 25 centimeters (10 inches), they are the largest eyes in the animal kingdom. They are positioned laterally on the head and are circular in outline. Each eye has an adjustable lens and a dark iris but no cornea.

The mouth is at the center of the circular crown of arms. Powerful chitinous jaws, encased in a muscular mass and capable of rotation and protrusion, are utilized to bite prey into chunks of a size suitable for swallowing. These parrotlike beaks, which can be more than 15 centimeters in length, consist of upper and lower mandibles. The strong upper mandible bears a pointed and acute-angled rostrum that forms a cutting edge with the lower mandible, which characteristically has a short, blunt rostrum and rounded winglike extensions. By careful examination the isolated jaws of individual squid species can be distinguished from those of other species. Hence it is possible to identify the otherwise unrecognizable digested contents of whale stomachs and the inclusions in ambergris, the waxy substance made by the sperm whale to enable it to purge itself of indigestible squid beaks.

The radula, or rasping tongue, is a characteristic feature of mollusks and usually consists of a long cuticular ribbon bearing transverse rows of chitinized teeth with cusps of various shapes. In *Architeuthis* the radula is small for an animal of such huge proportions. It is nonetheless impressive compared with the radulas of other mollusks, being about 100 millimeters long and a little more than 10 millimeters wide. Each row has a three-cusp central tooth and three smaller teeth on each side. Food bitten into chunks by the jaws is forced into the buccal cavity by the bolting movement of the radula. In addition many minute backward-sloping denticulations (the pharyngeal teeth) in the cuticle lining the pharynx facilitate swallowing and ensure that food moves inward to the alimentary canal.

The alimentary canal continues into a muscular esophagus that forces food by peristaltic contractions into the thick-walled stomach and the associated cecum. Digestive enzymes are secreted by the massive salivary glands and by the single medial "liver," or digestive gland, and the anterior pancreas. Assimilation is further advanced in the cecum. Wastes pass through the short intestine and out through a flapped rectum, which discharges near the internal opening of the funnel. The wastes are flushed out through the funnel with the water expelled for propulsion.

The squid obtains oxygen through a pair of long gills, which have many (sometimes more than 100) of the lamellae, or leaflets, that are the basic structures for the exchange of oxygen and carbon dioxide. Another feature is the ink sac characteristic of most squids. In the giant squid the sac is large and elongated, with a long duct that empties into the rectum. The black, mucous ink is thought to serve the giant squid as the ink of other squids does when it is extruded through the funnel in an escape reaction; that is, it maintains a cohesive shape resembling that of the squid, presumably confusing a predator as the squid jets away.

A female squid produces enormous numbers of whitish eggs. The eggs are relatively small, being from .5 millimeter to 1.4 millimeters long and .3 to .7 millimeter across. One specimen carried more than 5,000 grams (11 pounds), or perhaps a million eggs.

The males are externally differentiated from the females by their two hectocotylized arms. The testis consists of a white filamentous mass embedded posteriorly in the visceral mass. The spermatophores, long tubes filled with sperm, are manufactured in the complex spermatophoral apparatus, which is on the left side anterior to the testis.

The long, thin-walled spermatophoral sac (often called Needham's sac because it was first discovered and described by John Turberville Needham) with its basal seminal vesicle is attached to the visceral mass along the left side of the intestine in the mantle cavity. As such structures go it is enormous, reaching a length of a meter and serving as a storage chamber for hundreds (perhaps thousands) of spermatophores packed parallel to one another.

The distal extension of Needham's sac is often termed the penis, but that is probably a misnomer because the structure seems not to be an intromittent organ. Even in a small mature male it may be some 80 centimeters long and may protrude as much as 5.5 centimeters beyond the free edge of the mantle. The tip of the "penis" is mushroom-shaped and has in it a ventral slit about 15 millimeters long. Males apparently become sexually mature quite early; specimens with a mantle length of less than a meter have been found to have completely formed spermatophores.

Even though a giant squid is notably large and heavy out of the water, it is quite buoyant in the water. The buoyancy is due to the relatively high concentration of ammonium ions in the muscles of the mantle, head and arms. The concentration of ammonium ions probably explains why dead or dying squids rise to the surface and are often washed ashore. Ammonium ions have a specific gravity of 1.01, which is lower than that of seawater (1.022 at a depth of 50 meters and a temperature of 28 degrees Celsius). Without the ammonium ions the tissues of the squid are heavier than seawater, their average specific gravity being 1.046, but with them the animal can maintain its level in the water without having to expend energy by constant swimming.

A side effect of this characteristic is that a freshly stranded giant squid is made unpalatable by the strong, bitter taste of ammonia. In fact, the first observation that *Architeuthis* concentrates ammonium ions in certain muscles was made by three teuthologists, including one of us (Roper), who cooked a piece of giant squid for a party celebrating the completion of a doctoral examination. Subsequent analysis of the tissue confirmed the observation.

Although certain scholars have described *Architeuthis* as a strong swimmer, that is probably not the case. It is true that some oceanic squids are spectacular swimmers, but compared with them *Architeuthis* has a poorly developed musculature. In addition its fins are relatively small and weak, and the locking apparatus of the funnel is considerably less complex and powerful than the one in the strong swimmers.

Little is known about what the giant squid eats. Most of the specimens collected have an empty stomach; if the squid were not sick and "off its feed," it would have been unlikely to come to the surface or to be stranded. Even when something is found in a squid's stomach, the chances of identifying it are slim because the beak and the radula reduce the prey to small pieces and the digestive enzymes work fast. Since *Architeuthis* appears to be a relatively poor swimmer, it presumably is a somewhat passive and sluggish predator, unable to chase and capture large, active prey.

As for the predators of the giant squid, the principal one is the sperm, or cachalot, whale, *Physeter catodon*. Notwithstanding the limitations of *Architeuthis* as a swimmer, it is formidable in size and dexterity and has arms that are heavy, suckered and highly motile. The large eyes, representing an efficient detection system, and the ink sac are among the giant squid's protective mechanisms.

In the literature of giant squids one occasionally finds an account of a battle between a giant squid and a sperm whale at the surface of the sea. We must surmise that such a struggle is an attempt by the squid to escape from the whale rather than an attack on the whale by the squid. The scars of giant-squid suckers on the skin around the mouth and head of sperm whales attest to the reality of these battles.

Sperm whales prey on numerous kinds of fish, crustaceans, octopuses and squids, but much of their diet consists of *Architeuthis*. Although only a few giant-squid beaks may be found in a sperm whale's stomach along with hundreds of beaks from other squid species, the sheer size of a single giant squid may take up a third of the volume of a whale's stomach. The fact that a sperm whale's gut is usually found to contain large numbers of other prey but only one *Architeuthis* suggests that giant squids may be solitary animals, except possibly during the mating period.

Much has been written about the maximum size attained by *Architeuthis*, with assertions of total lengths exceeding 75 meters. There is no firm evidence for such assertions. The usual basis for them is the size of sucker scars on whales, but since a scar grows as a whale grows, it is unreliable evidence for the size of a squid unless it is demonstrably recent.

The largest specimen recorded in the scientific literature measured approximately 20 meters in total length. (It was stranded on a beach in New Zealand in 1880.) A significant part of this length, probably from 10 to 12 meters, consisted of the tentacles, which in a dead squid are notably elastic and easily stretched. In all other squid species the length of tentacles is always regarded as an imprecise component of measurement. The largest giant-squid mantle lengths known to us are in the range of five to six meters, the largest head lengths about one meter.

Even with the fairly large number of records of *Architeuthis* now available, it is still impossible to identify the precise habitat of this elusive animal. Most of the records come from strandings and from the stomach contents of sperm whales, neither of which give direct information about habitat. Sperm whales are known to feed in the depth range of 10 to 1,000 meters, and there is strong evidence that they go as deep as 2,000 meters. It is clear from their stomach contents that they feed partly along the bottom. Hence the *Architeuthis* individuals in sperm-whale stomachs could have been captured in midwater or on the bottom.

A fresh 12-foot section of a giant squid's tentacle was brought up 150 miles off the coast of California by a midwater trawl at a depth of 600 meters over a bottom depth of 4,000 meters. Sightings of giant squids swimming at the surface have been reported from off Newfoundland near the Grand Banks, where the water is less than 100 meters deep, and from the central Pacific over depths exceeding 4,000 meters. In recent years specimens of *Architeuthis* have been found in the vicinity of the Hawaiian Islands, where there is virtually no continental shelf and the bottom drops off sharply to several thousand meters.

Then there are the strandings. Many of the specimens are washed ashore dead, but enough are so fresh (or even barely alive) as to indicate that they were alive not too long before or too far from where they came ashore. Still, large squids, like large whales, may be alive when they are stranded but may also be far from their normal habitat because they are sick.

The fact that so few giant squids are captured in nets is intriguing, particularly now that huge midwater and bottom trawls are deployed by commercial fishermen and by research vessels. Do the squids detect the approach of a net and avoid capture, as so many other oceanic cephalopods do? Or do they live in habitats not normally entered by commercial fishermen or exploratory biologists because they are known to be unproductive, like the middle depths of the open ocean, or because they are too rocky and craggy and therefore dangerous to nets, like the deep canyons and edges of continental slopes? The fact that such questions still have to be asked about giant squids indicates how much remains to be learned about them.

Suggested Reading

Brodal, Alf, and Ragnar Fänge, ed. *The Biology of Myxine*. Universitetsforlaget, Oslo, 1963.

Coleman, Neville. *Australian Sea Life, South of 30 Degrees*. New York: Doubleday, 1987.

Gorbman, Aubrey et al. "The Hagfishery of Japan," *Fisheries* 15, no. 4 (July 1990): 12–18.

Jensen, David. "The Hagfish," *Scientific American* 214, no. 2 (February 1966): 82–90.

Jørgensen, J. M., J. P. Lomholt, R. E. Weber, and H. Malte. *The Biology of Hagfishes*. Chapman & Hall, 1998.

Kuiter, Rudie H. *Coastal Fishes of South Eastern Australia*. University of Hawaii Press, 1993.

Zahl, Paul. "Dragons of the Deep," *National Geographic* 153, no. 6 (June 1978): 838–845.

Beyond the Picnic Basket

The Army Ant

T. C. Schneirla and Gerard Piel

Wherever they pass, all the rest of the animal world is thrown into a state of alarm. They stream along the ground and climb to the summit of all the lower trees, searching every leaf to its apex. Where booty is plentiful, they concentrate all their forces upon it, the dense phalanx of shining and quickly moving bodies, as it spreads over the surface, looking like a flood of dark-red liquid. All soft-bodied and inactive insects fall an easy prey to them, and they tear their victims in pieces for facility in carriage. Then, gathering together again in marching order, onward they move, the margins of the phalanx spread out at times like a cloud of skirmishers from the flanks of an army.

That is how Henry Walter Bates, a Victorian naturalist, described the characteristic field maneuvers of a tribe of army ants. His language is charged with martial metaphor, but it presents with restraint a spectacle that other eyewitnesses have compared to the predatory expeditions of Genghis Khan and Attila the Hun.

Army ants abound in the tropical rain forests of Hispanic America, Africa and Asia. They are classified taxonomically into more than 200 species and distinguished as a group chiefly by their peculiar mode of operation. Organized in colonies 100,000 to 150,000 strong, they live off their environment by systematic plunder and pillage. They are true nomads, having no fixed abode. Their nest is a seething cylindrical cluster of themselves, ant hooked to ant, with queen and brood sequestered in a labyrinth of corridors and chambers within the ant mass. From these bivouacs they stream forth at dawn in tightly organized columns and swarms to raid the surrounding terrain. Their columns often advance as much as 35 meters an hour and may finally reach out 300 meters or more in an unbroken stream. For days at a time, they may keep their bivouacs fixed in a hollow tree or some other equally protected shelter. Then, for a restless period, they move on with every dusk. They swarm forth in a solemn, plodding procession, each ant holding to its place in line, its forward-directed antennae beating a hypnotic rhythm. At the rear come throngs of larvae-carriers and, at the very last, the big, wingless queen, buried under a melee of frenzied workers. Late at night they hang their new bivouac under a low-hanging branch or vine.

The army ant, observers are agreed, presents the most complex instance of organized mass behavior occurring regularly outside the homesite in any insect or, for that matter, in any subhuman animal. As such, it offers the student of animal psychology a subject rich in interest for itself. But it also provides an opportunity for original attack on some basic problems of psychology in general. The study here reported, covering the behavior of two of the Eciton species of army ants, was conducted by Schneirla over a 16-year period with extended field trips to the Biological Reservation on Barro Colorado Island in the Panama Canal Zone and to other ant haunts in Central America. In undertaking it, he had certain questions in mind. The central question, of course, was how such an essentially primitive creature as the ant manages such a highly organized and complex social existence. This bears on the more general consideration of organized group behavior as an adaptive device in natural selection.

The ant commends itself to study by man. Measured by the dispassionate standard of survival, it stands as one of the most successful of nature's inventions. It is the most numerous of all land animals both in number of individuals and number of species (more than 3,500 at present count). It has occupied the whole surface of the globe between the margins of eternal frost. Its teeming cities are to be found even on isolated atolls in mid-Pacific. The oldest of living orders, the ant dates back 60 million years to the early Jurassic period. More significant, the societies of ants probably evolved to their present state of perfection no less than 50 million years ago. Man, by contrast, is a dubious experiment in evolution that has barely got under way.

In the esteem of political philosophers, from Solomon to Winston Churchill, ants have shared honors with the two other classes of social insects, the bee and the termite. Of the three, the ant is by far the most various and interesting. Bees live in hives; termites burrow almost exclusively in wood. Ants are not so easily pigeonholed. Lord Avebury, a British formicologist, marveled at "the habits of ants, their large communities and elaborate habitations, their roadways, possession of domestic animals and, even, in some cases, of slaves!" He might have added that ants also cultivate agricultural crops and carry parasols. It is the social institutions of ants, however, that engender the greatest enthusiasm. Henry Christopher McCook, in his *Ant Communities and How They Are Governed, A Study in Natural Civics*, credited the ant with achieving the ultimate in democratic social order. The sight of an army ant bivouac put the British naturalist Thomas Belt in mind of Sir Thomas More's *Utopia*.

The marvels of ant life have led some thinkers into giddy speculation on the nature of ant intelligence. Few have put themselves so quaintly on record as Lord Avebury, who declared: "The mental powers of ants differ from those of men not so much in kind as in degree." He ranked them ahead of the anthropoid apes. Maurice Maeterlinck, author of *The Life of the Ant*, hedged: "After all, we have not been present at the deliberations of the workers and we know hardly anything of what happens in the depths of the formicary." Others have categorically explained ant behavior as if the creatures could reason, exchange information, take purposeful action and feel tender emotion. Describing a tribe of army ants on the march Paul Griswold Howes has "lieutenants keeping order or searching out the ground to be hunted or traveled next" and the privates in the line "obeying commands" and evincing "a wonderful sense of duty." R. C. Wroughton concluded from the precision of ant armies' maneuvers that "they are either the result of preconceived arrangement or are carried out by word of command."

Obviously anthropomorphism can explain little about ants, and it has largely disappeared from the current serious literature about ant behavior. Its place has been taken, however, by errors of a more sophisticated sort. One such is the concept of the "superorganism." This derives from a notion entertained by Plato and Aquinas that a social organization exhibits the attributes of a superior type of individual. Extended by certain modern biologists, the concept assumes that the biological organism, a society of cells, is the model for social organizations, whether ant or human. Plausible analogies are drawn between organisms and societies: division of function, internal communication, rhythmic periodicity of life processes and the common cycle of birth, growth, senescence and death. Pursuit of these analogies, according to the protagonists of the superorganism, will disclose that the same forces of natural selection have shaped the evolution of both organism and superorganism, and that the same fundamental laws govern their present existence.

This is of course a thoroughly attractive idea. It is representative, in the field of psychology, of current efforts in other fields of science to unify all observed facts by a single theory. But it possesses a weakness common to all Platonistic thinking. It erects a vague concept, "organism" or "organization," as an ultimate reality that defies explanation. The danger inherent in this arbitrary procedure is the bias that it imposes upon the investigator's approach to his problem. It reduces the gathering of evidence to the selection of appropriate illustrations and examples. This is a pitfall of which the investigator must be especially wary in the study of social behavior. Too often in this field theories and conclusions are composed of nine parts of rationalization to one part of evidence. The investigator in social science must be ruthless in discarding his preconceived notions, taking care to retain only the bare conceptual framework that is inductively supported by the already established evidence on his subject. In the gathering of new evidence he must impose on his work the same rules of repetition and control which prevail in the experimental sciences. Wherever possible he should subject his observations to experimental tests in the field and laboratory. In the area we are discussing this kind of work may at times seem more like a study of ants than an investigation of problems. But it yields more dependable data.

The contemporary ant, as will be shown, exhibits a comparatively limited capacity for learning. On the other hand, there is little that it needs to learn when it crawls out of the cocoon. By far the greater part of its behavior pattern is already written in its genes and represents the "learning" of its race, acquired many generations ago in the hard school of natural selection. The individual ant, as a matter of fact, is ill-equipped for advanced learning. By comparison with the sensitive perceptions of a human being, it is deaf and blind. Its hearing consists primarily in the perception of vibrations physically transmitted to it through the ground. In most species, its vision is limited to the discrimination of light and shadow. These deficiencies are partially compensated by the chemotactual perceptions of the ant, centered in its flitting antennae. Chiefly by means of its antennae, the ant tells friend from foe, locates its booty and, thanks to its habit of signing its trail with droplets from its anal gland, finds its way home to the nest.

In an investigation of ant learning, Schneirla found that individual ants are capable of significant feats of progress in a given situation, but that on the whole ant learning is by rote. His subject in this study was the common garden Formica ant, which is known to forage freely within a radius up to 75 meters around its nest. The learning situation was presented by a maze, interposed between the laboratory nest and feeding box, with maze passages open on the return route to the nest. Each individual ant was identified and followed by means of a number pasted on its gaster (abdomen). The ants betrayed no evidence of purposive behavior.

Control of clues provided in the maze by ant chemicals and variations in lighting revealed that the Formica ant possesses considerable learning capacity in its kinesthetic or "muscle" sense. Nevertheless, this study shows that the ant acquires merely a generalized maze habit, not an understanding of mazes. This conclusion is reinforced by another comparison with the rat. Confronted with abrupt changes in the maze layout, the rat will often exhibit plain evidence of emotional conflict, represented by an over-all deterioration of its learning progress. Ants, in the same situation, merely blunder ahead.

How the essentially uncomplicated repertory of the individual ant contrives, when ants act in concert, to yield the exceedingly complex behavior of the tribe is one of the most intricate paradoxes in nature. This riddle has been fruitfully explored under the guidance of the concept of "trophallaxis," originated by William Morton Wheeler. Trophallaxis (from the Greek *trophe*, meaning food, and *allaxis*, exchange) is based upon the familiar observation that ants live in biological thrall to their nestmates. Their powerful mutual attraction can be seen in the constant turning of one ant toward another, the endless antennal caresses, the licking and nuzzling. In these exchanges they can be seen trading intimate substances—regurgitated food and glandular secretions. Most ants are dependent for their lives upon this biosocial intercourse with their fellows. There is strong evidence that, as between larvae, workers and queen in a given tribe, there is an interchange of co-enzymes necessary to the existence of all. Army ant queens unfailingly sicken and die after a few days when isolated in captivity.

Trophallaxis, or "the spirit of the formicary," as Maeterlinck was pleased to call it, is therefore essentially chemical in nature. As can be seen by the mutual attractions and repulsions of ants for one another, their social chemicals are not only specific to species but also specific to colonies.

The well-established concept of trophallaxis naturally suggests that clues to the complex behavior of the ant armies should be sought in the relationships among individuals within the tribe.

The migratory habits of the ant armies follow a rhythmically punctual cycle. The *Eciton Hamatum* species, for example, wanders nomadically for a period of 17 days, then spends 19 or 20 days in fixed bivouac. This cycle coincides precisely with the reproductive cycle of the tribe. The army goes into bivouac when the larvae, hatched from the last clutch of eggs, have gone into the pupal state in their cocoons. At the end of the first week, the queen, with her gaster swollen to more than five times its normal volume, goes into a stupendous five- to seven-day labor in which she delivers 20,000 to 30,000 eggs. The daily foraging raids, which meanwhile have dwindled to a minimum, pick up again as the eggs hatch into a great mass of larvae. Then, on about the 20th day, the cocoons yield a new generation of callow workers, and the army sets off once more on its evening marches.

In determining this pattern of external social events Schneirla logged a dozen ant armies through one or more complete cycles, and upwards of 100 through partial cycles. The coinciding pattern of internal biological events was documented by brood samples taken from many different colonies at various stages in the reproductive cycle. Broods of more than 80 colonies were sampled, most of them repeatedly at intervals of a few days. In addition, detailed examinations were made of 62 queens in various phases of their physiological history and many of these were preserved for further study.

A sentimentalist considering this new picture of the army ant's domestic habits may find an explanation for its behavior more affecting than the food theory: the ants stay in fixed bivouac to protect the queen and her helpless young through the time when they are most vulnerable. Doubtless this is the adaptive significance of the process. But the motivation that carries 100,000 to 150,000 individual ants through this precisely timed cycle of group behavior is not familial love and duty but the trophallactic relationship among the members of the tribe. A cocooned and slumberous pupa, for example, exerts a quieting influence upon the worker that clutches it in its mandible—somewhat as a thumb in the mouth pacifies an infant. But as it approaches maturity and quickens within its cocoon, the pupa produces precisely the reverse effect. Its stirring and twitching excite the workers to pick up the cocoon and snatch it from one another. As an incidental result, this manhandling effects the delivery of the cocoon's occupant. (Cocoons in which the pupae were killed by needle excited no such interest among the workers and remained unopened.)

The stimulus of the emerging brood is evident in a rising crescendo of excitement that seizes the whole community. Raiding operations increase in tempo as the hyperactive, newly delivered workers swarm out into the marching columns. After a day or two, the colony stages an exceptionally vigorous raid that ends in a night march. The bivouac site is left littered with empty cocoons. Later in the nomadic period, as the stimulus of the callow workers wanes, the larvae of the next generation become the source of colony "drive." Fat and squirming, as big as an average worker, they establish an active trophallactic relationship with the rest of the tribe. Workers constantly stroke them with their antennae, lick them with their mouth parts and carry them bodily from place to place. Since the larvae at this stage are usually well distributed throughout the corridors and the chambers of the overnight bivouac, their stimulus reaches directly a large number of the workers. This is reflected in the sustained vigor of the daily raids, which continue until the larvae spin their cocoons.

These observations are supported by a variety of experimental findings in the field and laboratory. The role of the callow workers in initiating the movement to break bivouac was confirmed by depriving a number of colonies of their callow broods. Invariably, the raiding operations of the colony failed to recover from the lethargic state characteristic of the statary periods. Some tribes even extended their stay in fixed bivouac until the larvae grew large and active enough to excite the necessary pitch of activity. To test the role of the larval brood, captured tribes were divided into part-colonies of comparable size. The group with larvae showed much greater activity than those that had no larvae or that had cocoons in the early pupal state.

The interrelationships among members of the colony thus provide a complete explanation for the behavior cycle of the army ant. It should be observed, in conclusion, that the whole complex process is carried out by individuals that do not themselves originate the basic motivations of their behavior.

Long before the intricacies of its domestic existence were suspected, the army ant's reputation as a social animal was firmly established by its martial conduct in external affairs. It does not require an overactive imagination to perceive the classic doctrines of offensive warfare spelled out by the action of an ant army in the field. It carries through the maneuvers of wheeling, flanking and envelopment with a ruthless precision. But to find its motivations and explain its mechanics, one must consult the ant, not von Clausewitz.

Army ant raids fall into one of two major patterns. They are organized either in dense swarms which form at the head of the column or in a delicate tracery of capillary columns branching out at the forward end of the main raiding column. Both types of raiding are found in subgenera of each of the common species of Central American army ant. The *Eciton eciton* species was selected for this study because it leads its life entirely on the surface of the jungle floor and is thus accessible to continuous observation. Whether the army ants raid in swarm or column, however, the essential mechanics of their behavior are substantially the same.

The bivouac awakes in the early dawn. The stir of activity mounts steadily as the light increases. In strands and clusters, the workers tumble out of the bivouac into a churning throng on the ground. A crowding pressure builds up within this throng until, channeled by the path of least resistance, a raiding column suddenly bursts forth. The ants in the column are oriented rigidly along the line of travel blazed by the chemical trail of the leaders. The minims and medium-size workers move in tight files in the center. The "workers major," displaced by the unstable footing afforded by the backs of their smaller fellows, travel along each side. This arrangement no doubt lends suggestive support to the major's legendary role of command. It has an adaptive significance in that it places the biggest and most formidable of the workers on the flanks. Unless disturbed, however, the majors hug the column as slavishly as the rest. The critical role of the tribal chemical in creating this drill sergeant's picture of order may be demonstrated by a simple field experiment. Removal of the chemically saturated litter from the trail brings the column to an abrupt halt. A traffic jam of ants piles up on the bivouac side of the break and is not relieved until enough ants have been pushed forward to reestablish the chemical trail.

The path of an ant army, whether in swarms or columns, shows no evidence of leadership. On the contrary, each individual makes substantially the same contribution to the group behavior pattern. The army's course is directed by such wholly chance factors as the stimulus of booty and the character of the terrain. On close inspection, therefore, it appears that the field operations of ant armies approximate the principles of hydraulics even more closely than those of military tactics. This impression is confirmed by analysis of the flanking maneuver as executed by the swarm raiders. A shimmering pattern of whirls, eddies and momentarily milling vortices of ants, the swarm advances with a peculiar rocking motion. First one and then the other end of the elliptical swarm surges forward. This action results in the outflanking of quarry, which is swiftly engulfed in the overriding horde of ants. It arises primarily, however, from an interplay of forces within the swarm. The stimulus of booty will accelerate the advance of a flank. The capture of booty will halt it and bring ants stampeding in for a large-scale mopping-up party. But raiding activity as such is only incidental to the process. Its essential character is determined by the stereotyped behavior of the individual ant with its limited repertory of responses to external stimuli.

The profoundly simple nature of the beast is betrayed by an ironic catastrophe that occasionally overtakes a troop of army ants. It can happen only under certain very special conditions. But when these are present, army ants are literally fated to organize themselves in a circular column and march themselves to death. Postmortem evidence of this phenomenon has been found in nature; it may be arranged at will in the laboratory. Schneirla has had the good fortune to observe one such spectacle in nature almost from its inception to the bitter end.

The ants, numbering about 1,000, were discovered at 7:30 a.m. on a broad concrete sidewalk on the grounds of the Barro Colorado laboratories. They had apparently been caught by a cloudburst that washed away all traces of their colony trail. When first observed, most of the ants were gathered in a central cluster, with only a company or two plodding, counterclockwise, in a circle around the periphery. By noon all the ants had joined the mill, which had now attained the diameter of a phonograph record and was rotating somewhat eccentrically at fair speed. At 10:00 p.m. the mill was found divided into two smaller counterclockwise spinning discs. At dawn the next day the scene of action was strewn with dead and dying Ecitons. A scant three dozen survivors were still trekking in a ragged circle. By 7:30, 24 hours after the mill was first observed, the various small myremicine and dolichoderine ants of the neighborhood were busy carting away the corpses.

The general mechanics of the mill are fairly obvious. The circular track represents the vector of the individual ant's centrifugal impulse to resume the march and the centripetal force of trophallaxis that binds it to its group. Where no obstructions disturb the geometry of these forces, as in the artificial environment of the laboratory nest or of a sidewalk, the organization of a suicide mill is almost inevitable. Fortunately for the army ant, it is rare in the heterogeneous environment of nature.

The army ant suicide mill provides an excellent occasion for considering the comparative nature of social behavior and organization at the various levels from ants to men. Other animals occasionally give themselves over to analogous types of mass action. Circular mills are common among schools of herring. Stampeding cattle, sheep jumping fences blindly in column and other instances of pell-mell surging by a horde of animals are familiar phenomena.

We are required, however, to study the relationship of the pattern to other factors of individual and group behavior in the same species. In the case of the army ant, of course, the circular column really typifies the animal. Among mammals, such simplified mass behavior occupies a clearly subordinate role. Their group activity patterns are chiefly characterized by great plasticity and capacity to adjust to new situations. This observation applies with special force to the social potentialities of man.

The same reservations apply to the analogies cited to support the superorganism theory. Consider, for example, the analogy of "communication." Among ants it is limited to the stimulus of physical contact. One excited ant can stir a swarm into equal excitement. But this behavior resembles the action of a row of dominoes more than it does the communication of information from man to man. The difference in the two kinds of "communication" requires two entirely different conceptual schemes and preferably two different words.

As for "specialization of functions," that is determined in insect societies by specialization in the biological makeup of individuals. Mankind, in contrast, is biologically uniform and homogeneous. Distinctions among men are drawn on a psychological basis. They break down constantly before the energies and talents of particular individuals.

Finally, the concept of "organization" itself, as it is used by the superorganism theorists, obscures a critical distinction between the societies of ants and men. The social organizations of insects are fixed and transmitted by heredity. But members of each generation of men may, by exercise of the cerebral cortex, increase, change and even displace given aspects of their social heritage. This is a distinction that has high ethical value for men when they are moved to examine the conditions of their existence.

Slavery in Ants

Edward O. Wilson

The institution of slavery is not unique to human societies. No fewer than 35 species of ants, constituting six independently evolved groups, depend at least to some extent on slave labor for their existence. The techniques by which they raid other ant colonies to strengthen their labor force rank among the most sophisticated behavior patterns found anywhere in the insect world. Most of the slave-making ant species are so specialized as raiders that they starve to death if they are deprived of their slaves. Together they display an evolutionary descent that begins with casual raiding by otherwise free-living colonies, passes through the development of full-blown warrior societies and ends with a degeneration so advanced that the workers can no longer even conduct raids.

Slavery in ants differs from slavery in human societies in one key respect: the ant slaves are always members of other completely free-living species that themselves do not take slaves. In this regard the ant slaves perhaps more closely resemble domestic animals—except that the slaves are not allowed to reproduce and they are equal or superior to their captors in social organization.

The famous Amazon ants of the genus *Polyergus* are excellent examples of advanced slave makers. The workers are strongly specialized for fighting. Their mandibles, which are shaped like miniature sabers, are ideally suited for puncturing the bodies of other ants but are poorly suited for any of the routine tasks that occupy ordinary ant workers. Indeed, when *Polyergus* ants are in their home nest their only activities are begging food from their slaves and cleaning themselves ("burnishing their ruddy armor," as William Morton Wheeler once put it).

When *Polyergus* ants launch a raid, however, they are completely transformed. They swarm out of the nest in a solid phalanx and march swiftly and directly to a nest of the slave species. They destroy the resisting defenders by puncturing their bodies and then seize and carry off the cocoons containing the pupae of worker ants.

When the captured pupae hatch, the workers that emerge accept their captors as sisters; they make no distinction between their genetic siblings and the *Polyergus* ants. The workers launch into the round of tasks for which they have been genetically programmed, with the slave makers being the incidental beneficiaries. Since the slaves are members of the worker caste, they cannot reproduce. In order to maintain an adequate labor force, the slave-making ants must periodically conduct additional raids.

The slave raiders obey what is often called Emery's rule. In 1909 Carlo Emery noted that each species of parasitic ant is genetically relatively close to the species it victimizes. This relation can be profitably explored for the clues it provides to the origin of slave making in the evolution of ants. The territorial defense rather than food is the evolutionary prime mover. I brought together different species of *Leptothorax* ants that normally do not depend on slave labor. When colonies were placed closer together than they are found in nature, the larger colonies attacked the smaller ones and drove away or killed the queens and workers. The attackers carried captured pupae back to their own nests. The pupae were then allowed by their captors to develop into workers. In the cases where the newly emerged workers belonged to the same species, they were allowed to remain as active members of the colony. When they belonged to a different *Leptothorax* species, however, they were executed in a matter of hours. One can easily imagine the origin of slave making by the simple extension of this territorial behavior to include tolerance of the workers of related species. The more closely related the raiders and their captives are, the more likely they are to be compatible. The result would be in agreement with Emery's rule.

One species that appears to have just crossed the threshold to slave making is *Leptothorax duloticus*, a rare ant that so far has been found only in certain localities in Ohio, Michigan and Ontario. The anatomy of the worker caste is only slightly modified for slave-making behavior, suggesting that in evolutionary terms the species may have taken up its parasitic way of life rather recently.

In experiments with laboratory colonies I was able to measure the degree of behavioral degeneration that has taken place in *L. duloticus*. Like the Amazon ants, the *duloticus* workers are highly efficient at raiding and fighting. When colonies of other *Leptothorax* species were placed near a *duloticus* nest, the workers launched intense attacks until all the pupae of the other species had been captured.

In the home nest the *duloticus* workers were inactive, leaving almost all the ordinary work to their captives. When the slaves were temporarily taken away from them, the workers displayed a dramatic expansion in activity, rapidly taking over most of the tasks formerly carried out by the slaves. The *duloticus* workers thus retain a latent capacity for working, a capacity that is totally lacking in more advanced species of slave-making ants.

The *duloticus* workers that had lost their slaves did not, however, perform their tasks well. Their larvae were fed at infrequent intervals and were not groomed properly, nest materials were carried about aimlessly and were never placed in the correct positions, and an inordinate amount of time was spent collecting and sharing diluted honey. More important, the slaveless ants lacked one behavior pattern that is essential for the survival of the colony: foraging for dead insects and other solid food. They even ignored food placed in their path. When the colony began to display signs of starvation and deterioration, I returned to them some slaves of the species *Leptothorax curvispinosus*. The bustling slave workers soon put the nest back in good order, and the slave makers just as quickly lapsed into their usual indolent ways.

Not all slave-making ants depend on brute force to overpower their victims. Quite by accident Fred E. Regnier and I discovered that some species have a subtler strategy. While surveying chemical substances used by ants to communicate alarm and to defend their nest, we encountered two slave-making species whose substances differ drastically from those of all other ants examined so far. These ants, *Formica subintegra* and *Formica pergandei*, produce remarkably large quantities of decyl, dodecyl and tetradecyl acetates. Further investigation of *F. subintegra* revealed that the substances are sprayed at resisting ants during slave-making raids. The acetates attract more invading slave makers, thereby serving to assemble these ants in places where fighting breaks out. Simultaneously the sprayed acetates throw the resisting ants into a panic. Indeed, the acetates are exceptionally powerful and persistent alarm substances. They imitate the compound undecane and other scents found in slave species of *Formica*, which release these substances in order to alert their nestmates to danger. The acetates broadcast by the slave makers are so much stronger, however, that they have a long-lasting disruptive effect. For this reason Regnier and I named them "propaganda substances."

We believe we have explained an odd fact first noted by Pierre Huber 191 years ago in his pioneering study of the European slave-making ant *Formica sanguinea*. He found that when a colony was attacked by these slave makers, the survivors of the attacked colony were reluctant to stay in the same neighborhood even when suitable alternative nest sites were scarce.

Regnier and I were further able to gain a strong clue to the initial organization of slave-making raids. We had made a guess, based on knowledge of the foraging techniques of other kinds of ants, that scout workers direct their nestmates to newly discovered slave colonies by means of odor trails laid from the target back to the home nest. In order to test this hypothesis we made extracts of the bodies of *F. subintegra* and of *Formica rubicunda*, a second species that conducts frequent, well-organized raids through much of the summer. Then at the time of day when raids are normally made we laid artificial odor trails, using a narrow paintbrush dipped in the extracts we had obtained from the ants. The trails were traced from the entrances of the nest to arbitrarily selected points one or two meters away.

The results were dramatic. Many of the slave-making workers rushed forth, ran the length of the trails and then milled around in confusion at the end. When we placed portions of colonies of the slave species *Formica subsericea* at the end of some of the trails, the slave makers proceeded to conduct the raid in a manner that was apparently the same in every respect as the raids initiated by trails laid by their own scouts.

The evolution of social parasitism in ants works like a ratchet, allowing a species to slip further down in parasitic dependence but not back up toward its original free-living existence. An example of nearly complete behavioral degeneration is found in one species of the genus *Strongylognathus*, which is found in Asia and Europe. Most species in this genus conduct aggressive slave-making raids. They have characteristic saber-shaped mandibles for killing other ants. The species *Strongylognathus testaceus*, however, has lost its warrior habits. Although these ants still have the distinctive mandibles of their genus, they do not conduct slave-making raids. Instead an *S. testaceus* queen moves into the nest of a slave-ant species and lives alongside the queen of the slave species. Each queen lays eggs that develop into workers, but the *S. testaceus* offspring do no work. They are fed by workers of the slave species.

Thus *S. testaceus* is no longer a real slave maker. It has become an advanced social parasite of a kind that commonly infests other ant groups. For example, many species of ant play host to parasites such as beetles, wasps and flies, feeding them and sheltering them. Ant slavery is a genetic adaptation found in particular species that cannot be judged to be more or less successful than their non-slave-making counterparts. The slave-making ants offer a clear and interesting case of behavioral evolution.

The Fire Ant

Edward O. Wilson

"Fire ant" is the common name of many ant species distributed throughout the tropical and warm temperate regions of the New World. The sting of these ants causes a burning sensation, hence the name. Three species of fire ant are native to the southern U. S.; a fourth (*Solenopsis saevissima*) was introduced from South America around 1918. For 10 years the imported fire ant lived within the city limits of Mobile, Alabama; then it began to spread.

It is not only a serious pest but a versatile one. In South America the normal diet of the fire ant appears to consist mostly of seeds, the flesh of insects and "honeydew" gathered from living insects such as aphids. But its dense populations in the U. S. have extended this diet to include, to the grief of farmers, the seedlings of several important food crops and the newborn young of poultry and livestock. Nesting fire ants build large mounds, numbering up to 50 an acre, which hamper plowing and harvesting. Moreover, worker ants swarm aggressively out of the nests at the slightest disturbance, making manual labor in infested fields painful and difficult. Some farmers who have heavily infested land are unable to hire sufficient help, and are forced to abandon land to the ants. The total agricultural loss in the South probably extends into millions of dollars.

This unhappy tale has a special interest for the student of organic evolution. A fundamental problem of evolution is: How does a species adapt itself to a new environment? The problem is dramatized by the special case of animals and plants transported by man to areas far from their native ranges. Some transplanted species are immediately successful, building up huge populations until they become dominant members of the local fauna and flora; related species completely fail to establish themselves. Still other species follow a more baffling pattern. For a time they maintain a limited and precarious "beachhead"; then suddenly and unpredictably they explode into a phase of rapid expansion. The imported fire ant occupies the last category.

Like most other ants, fire ants are dispersed by the nuptial flights of males and winged virgin queens. During the flight a queen may travel as far as five miles from the colony of her origin. After she mates with the male she descends to earth, excavates a simple burrow in the soil, and over a period of several days lays approximately 100 eggs. The first brood of worker ants then develops very rapidly. The eggs hatch in about nine days, the larvae change to pupae after about the same period, and adults emerge from the pupae about a week later. Often small groups of queens cooperate to found colonies. Later, however, the first emerging workers execute the surplus queens so that only one remains. Once established, the young colony grows with startling speed. Within four or five months it contains over 1,000 workers. In a year it seethes with tens of thousands of workers and has reached sexual maturity; that is, it has begun to produce the winged males and queens which start the life cycles of new colonies.

Given this means of dispersion, exactly how did the imported fire ant spread in the U. S.? As I have indicated, the imported population was at first quiescent and then exploded. We now know that at the beginning of the explosive phase an important change in the genetic structure of the population occurred. In the 1920s, following the introduction of the colony into Mobile, it consisted entirely of a relatively large, blackish-brown form that corresponded exactly to the southernmost race of the mother population in South America. The range of this race is northern and central Argentina and part of southern Uruguay. The founding colony (or colonies) may have come from Buenos Aires or Montevideo in ships. Once established in Mobile, the dark form was not notably successful. William S. Creighton, in 1928, found it limited to Mobile and the suburban community of Spring Hill. There was no inkling of the explosion to come.

Sometime in the 1930s a second form of the imported fire ant made its appearance in the Mobile area. It was reddish-brown in color, smaller in size than the original immigrant, and it built smaller nests. Its origin is not positively known. Three possibilities have been considered: (1) that the second form was introduced from another part of South America, (2) that it was a mutation of the original dark form, (3) that it represented a recombination of genes already present in the dark form. Several lines of evidence point to the first alternative. Where light and dark colonies meet we find colonies of many intermediate colors. This suggests that the variation in color is controlled not by one gene but by several. Moreover, the light and dark forms differ in characteristics other than color, and these characteristics vary independently of one another; thus they are probably controlled by different groups of genes. So it seems unlikely that mutation or recombination can account for the sudden appearance of the complex genetic structure of the light form.

Furthermore, the light form, or its close equivalent, occurs abundantly in certain parts of northern Argentina and southern Bolivia. Perhaps like the dark form it was introduced into the U.S. in cargo shipped out of Buenos Aires or Montevideo. Whatever the origin of the light form, the significant fact is that it appeared at just about the time that the population as a whole began its rapid growth in all directions out of Mobile.

When I first began to study the imported fire ant in 1949, I noticed that the two color forms were distributed around Mobile in a curious and significant pattern. The dark form was very rare within the city itself, despite the fact that it had been the predominant or exclusive form when Creighton studied it there 20 years earlier. Now it was limited to a few isolated localities concentrated mostly along the southern periphery of the range. Everywhere else in the Mobile area the teeming colonies were composed of the light form. To the north, in the Mississippi towns of Meridian and Artesia, there were small secondary populations consisting entirely of the dark form. Investigation showed that these localities had been colonized by ants from the Mobile area during the 1930s, probably while the dark form still predominated in the primary population. Other isolated populations were found at Thomasville and Selma in Alabama. These consisted entirely of the light form. As one might have predicted, it was subsequently disclosed that these light-form populations were quite recent in origin, being no more than five years old in 1949. Compared to the dark-form populations of Missis-

sippi, they were remarkably successful and fast-growing. The Selma population, in fact, already exceeded in size both Mississippi populations taken
together.

All this information clearly indicated that the imported fire ant was in
the midst of a rapid evolutionary change. The conclusion seemed
inescapable that the light form, which had originated in the Mobile area
sometime after the introduction of the dark form, was adaptively superior
to the dark form and was replacing it over most of its range.

The future of the imported fire ant in the U.S. is difficult to predict.
Energized by the highly adaptive genes of the light form, its rapidly growing
populations may come to cover all the southeastern states. As a warm tem-
perate-zone species that nests exclusively outdoors, it may never succeed in
pushing north much beyond Alabama and North Carolina, but within this
range it will undoubtedly continue to wax as one of the most noxious of all
insect pests.

It is clear that long-range control of such vast insect populations is going
to be a very complicated matter. It is likely that biological control—the
introduction of new parasites and predators that attack the fire ant—will
have to be attempted. This technique, which has been so effective in stop-
ping other insect pests, has not yet been investigated with respect to the fire
ant. Finally there is the ironic fact that the fire-ant infestation, if left alone
long enough, would probably abate by itself. A general characteristic of
introduced populations is that in time native elements adjust to them and
eventually reduce their density. But the period of adjustment could take
years or decades, and continued research on control measures seems economically imperative. Meantime the imported fire ant will provide valuable
clues as to the kind of genetic processes that underlie the adaptation of animal species to new environments.

Singing Caterpillars, Ants and Symbiosis

Philip J. DeVries

As anyone who has picnicked atop an anthill can attest, ants vigorously defend the food they find and strongly discourage poaching on their territory by other species (human or otherwise). Yet some insects, including a breathtaking variety of butterfly caterpillars, can not only trespass safely but enter into mutually beneficial partnerships with ants. Such relationships are intriguing examples of symbiosis, in which two or more species may live and interact intimately. Because symbioses offer great clues for understanding the complex interactions between multiple species, they are of special interest to evolutionary ecologists, who are concerned with how and why organisms have evolved their particular traits and behaviors.

The ability to form symbioses with ants and to exploit their pugnacious characteristics is well known in two major groups of organisms: plants and herbivorous insects belonging to the orders Homoptera (aphids, cicadas and related insects) and Lepidoptera (butterflies, skippers and moths). For all these organisms, the presentation of food in the form of tasty secretions seems to be fundamental to maintaining the association with the ants. Plants may provide secretions through extrafloral nectaries on their leaves. Ants drawn to these nectaries defend the plants against herbivorous insects. Similarly, insects can offer secretions to ants through specialized organs.

Myrmecophily, the ability to associate symbiotically with ants, has evolved among only two families of butterflies: the Lycaenidae (which are known as hairstreaks and blues and are found everywhere in the world) and the Riodinidae (or metalmarks, which are found almost exclusively in the American tropics). Together these families make up what are commonly called the lycaenoid butterflies. Lycaenoids are seldom noticed—they have wingspans of less than two inches—but they constitute 40 percent of the more than 13,500 known butterfly species and encompass a staggering diversity of colors and patterns. As delightful to the eye as lycaenoid butter-flies may be, it is their caterpillars that dazzle the imagination of evolution-ary biologists because many of them have specialized organs that mediate their symbioses with ants.

Researchers have shown that the symbioses between lycaenid caterpillars and ants can range from the mutualistic, in which both species benefit, to the parasitic, in which one benefits at the expense of the other. In some instances, the symbioses lead to complicated life cycles for one or both species.

Until recently, the only detailed work on symbioses between riodinid caterpillars and ants was that done years ago by Gary N. Ross. For a very long time, therefore, our grasp of the evolution and ecology of butterfly-ant symbioses was based almost entirely on studies of the family Lycaenidae. Research on riodinid caterpillars, however, yielded comparative informa-tion that has helped reinterpret the evolution and consequences of symbi-otic associations with ants.

My interest in caterpillar-ant interactions started some years ago in Brunei, Borneo, where I first observed interactions between ants and a lycaenoid butterfly. At the time, most of my work centered on other aspects of butterfly biology, but from this incidental observation grew a fascination with myrmecophilous (ant-loving) caterpillars.

The riodinid butterfly that I have studied in greatest detail is *Thisbe ire-nea*, which lives in a variety of tropical forest habitats between Mexico and Brazil. It is typical of riodinid caterpillars that form symbioses with ants. A female *Thisbe* deposits individual eggs on sapling trees of the genus *Croton*; after caterpillars emerge from these eggs, they feed on the tree. Ants are drawn to *Croton* because at the base of each leaf is an extrafloral nectary. (Such nectaries are common features of many tropical plants.) The ants patrolling the trees are the same ones with which the caterpillars form sym-bioses.

When I began my study of *Thisbe* in Barro Colorado, Panama, all that was known about the caterpillars was that they fed on *Croton* and were typically found in association with ants. The logical first step was to see what would happen to the caterpillars without ants. I therefore removed all the insects from several populations of *Croton* saplings and then smeared the base of each plant with a sticky resin. This arrangement would deny all crawling insects, such as ants, access to the trees but would not prevent female butterflies from laying eggs there. On half the trees, however, I placed a stick bridge over the resin to allow ants to return.

For the next 10 months, I conducted a weekly census of ants and caterpillars on all the trees. The results showed that the trees with ants on them accrued more caterpillars than did the trees from which ants were excluded. A likely explanation for this pattern seemed to be that on the plants lacking ants, flying predators were removing the caterpillars.

Of the natural predators that caterpillars face, the social wasps are especially significant, particularly in the tropics. Social wasps spend a large proportion of their adult lives searching vegetation for caterpillars. When a female wasp finds a caterpillar, she kills it with her sting, cuts up the body and carries the meat back to her nest to feed the hungry wasp grubs.

To test whether ants protected caterpillars from wasps, I placed two potted plants in an area where wasp nests were abundant and allowed ants to crawl on only one of the plants. I then placed a caterpillar on each plant and timed how long it survived. As eventually became obvious, in the absence of ants, caterpillars did not remain on plants for very long: often within minutes, wasps killed the caterpillars and carried them away. In contrast, if ants were present, they vigorously defended the caterpillars against wasp attacks. Wasp predation therefore explained why I found fewer caterpillars on the plants from which ants had been excluded.

These simple experiments showed that ants performed a valuable service to *Thisbe* caterpillars by protecting them from predators. That observation raised a further question about the symbiosis: what did the ants gain in return for their efforts? Or to put it another way, how do the caterpillars induce ants to engage in a risky defense of an organism that is not part of their colony?

The answer rests, in part, with a suite of specialized organs found on the caterpillar. Ants generally ignore young *Thisbe* caterpillars that are still in their first and second instars, or developmental stages.

Caterpillars in their third and later instars have three sets of so-called ant organs that are important for maintaining the tending activities by the ants. The most noticeable are a pair of extrudable glands called nectary organs. Located on the caterpillar's posterior segments, the nectary organs look like the fingers of a rubber surgical glove. When an ant stokes the posterior area of a caterpillar with its antennae, these organs extrude from the body and secrete a drop of clear fluid at their tips that the ant drinks eagerly. The organs then withdraw back into the body. The ants, however, are so captivated by the secretion that they stroke the caterpillar relentlessly to solicit it. I estimate that the ants attending a *Thisbe* caterpillar seek more of the secretion at least once every minute.

These same ants also obtain secretions from the extrafloral nectaries of *Croton* trees, but they seem to prefer tending the caterpillars to tending the nearby plant nectaries. The caterpillar secretions are quite different from those offered by the *Croton* plants. In effect, the secretion of the caterpillars is haute cuisine. True, the extrafloral nectar is a 33 percent mixture of various sugars, whereas the caterpillar secretion has almost no detectable sugar, but the secretion contains much higher concentrations of amino acids. Ants can obtain a meal from the caterpillars that is more nutritious than what they can get from the plant nectaries, even if it is not so sweet.

Studies of the nectary organs solved only part of the caterpillar-ant symbiosis puzzle. Ants are selfless automatons that provide food, defense and brood care for their colony as a whole. Yet individual ants sometimes stay with *Thisbe* caterpillars for a week or more. Instead of spending time protecting members of a different species, why don't the attending ants immediately return to their nest with the caterpillar secretion, as they would with plant nectar or other delicacies?

Other ant organs on the caterpillars bear on that question. The caterpillars of *Thisbe* and other riodinid species have a pair of tentacle organs, brush-tipped glands just behind the head, that appear to exert a chemical influence on ant behavior. When these organs emerge from the body, the attending ants near the caterpillar almost immediately snap into a defensive posture, their mandibles agape, abdomen curled under the body. I found that when the ants assumed this posture, moving a small piece of wood or thread near the caterpillar precipitated an aggressive attack: the ants would rush at the object, bite it and attempt to sting it.

These observations of *Thisbe* and other riodinid species suggest that the tentacle organs discharge some chemical similar to an ant alarm pheromone—a substance that the ants would use among themselves to signal an attack on the colony. Unfortunately, the chemical nature of the emissions from the tentacle organs is still unknown, so it is impossible to say how they compare with actual ant alarm pheromones. It does seem clear, however, that the function of these organs is to seize the attention of ants and to help keep it focused on the caterpillars.

A third type of ant organ was discovered in 1926 by entomologist Carlos T. Bruch of Buenos Aires, who provided the first description of a myrmecophilous riodinid species found in Argentina. In addition to the extrudable nectary and tentacle organs, Bruch found a pair of tiny, movable, rodlike appendages that sprang from the front of the first thoracic segment and projected over the head. Such appendages were then unknown on any other caterpillar.

Forty years later in Mexico, Ross described similar appendages on a different riodinid species. Ross named these appendages vibratory papillae and suggested that their motion might convey vibrations to ants. It is now known that most myrmecophilous riodinid caterpillars have vibratory papillae as ant organs.

My observations indicated that the vibratory papillae on *Thisbe* caterpillars functioned as Bruch and Ross had described. I also noticed, however, that while the vibratory papillae were in motion, the caterpillars moved their heads in and out. I was struck by the similarity of these head movements to those of long-horned beetles, which produce an audible sound when they move their heads in and out. (Predators seem to drop the beetles when they hear the squeaks.) Even though I could hear nothing coming from *Thisbe* caterpillars, I became convinced that they were producing sounds.

I examined the structures with a scanning electron microscope. Under magnification, the vibratory papillae clearly showed sharp concentric rings along their shafts. The top of the head, which the papillae touched when they vibrated, bore small bumps or granulations that looked like guitar picks. Side by side in those micrographs, the vibratory papillae and the head were strongly reminiscent of a guiro, a Latin American percussion instrument that is played by sliding a wooden stick across the grooves in a carved gourd. Thus, both the morphology and the behavior of the caterpillar suggested that these papillae and the head acted jointly as a sound-producing organ.

No one had ever heard a caterpillar make a sound, however, so this crazy idea needed to be tested. I consequently returned to Panama with an exceptionally sensitive microphone and amplifier. On the very day of my arrival, I discovered that the instruments did indeed allow me to hear and record calls from the caterpillars. Their low-amplitude calls were audible only when the microphone was touching the body of a caterpillar or the surface on which it stood. The calls therefore travel through the solid substrate rather than through the air, which partially explained why no one had ever heard caterpillar calls during normal observation.

By making recordings of individual caterpillars and then removing their vibratory papillae, I learned that the caterpillars could produce calls as long as they had at least one papilla. As analyses of the recordings revealed, the calls were strongest when two vibratory papillae were present and weaker by almost half when one was removed. Caterpillars without any vibratory papillae were completely mute even though they continued to move their heads. Because these papillae are replaced when a caterpillar sheds its skin, the mute caterpillars regained their "voices" when they molted to the next instar. (That fact was experimentally useful because individual caterpillars could be used several times, and they acted as built-in controls.) Taken together, these observations verified that the vibratory papillae, the head granulations and the head motions worked literally in concert as components of a call-producing system.

Just as the chemical stimulus from the tentacle organs seems to mimic a pheromonal signal among ants, the calls of the caterpillars seem to parallel the auditory communications of ants, at least in some respects. When ants find a food resource or when they are alarmed, they create vibrations that travel through the substrate on which they stand and attract their nest mates. Many varieties of ant produce vibrations by tapping their abdomen on the substrate, but others have well-developed sound-making organs. The

Vibratory papillae extending over the caterpillar's head from the first segment of the thorax help the insect to produce an acoustic call that attracts ants. Each papilla is a rod covered by concentric grooves. When the caterpillar moves its head in and out, microscopic granulations, or bumps, on the head slide across the grooves, producing vibrations.

calls of ants and caterpillars have approximately the same frequencies and pulse rates.

These similarities suggest that the ability of caterpillars to call and the characteristics of the calls themselves evolved through natural selection by ants. In other words, the ants determined which caterpillars survived by ignoring strange calls and responding to antlike calls.

That behavior is noteworthy because it violates what biologists have generally understood about insect communication systems—that calls evolve in response to direct selection for sexual or defensive traits. For example, the chirping of a male cricket summons potential mates and warns other males away from its territory. Vigorous males proliferate faster than do their sluggish brothers because their calls attract larger numbers of receptive females. The calling ability of crickets and the characteristics of their calls thus evolve directly in response to selection by females of the same species.

In caterpillar and ant symbioses, however, the calls produced by one species have evolved in response to selection by an unrelated species. The evolution of the caterpillars has been guided by selection acting on completely different characteristics in ants. The caterpillar-ant system therefore provides a new arena for the study of insect communication, particularly in other insect species that form symbioses with ants.

Caterpillar-ant symbioses are typically facultative rather than obligate— that is, the two species benefit each other but neither is absolutely dependent on the other for survival. Indeed, the involvement of species in a symbiosis can be quite flexible: depending on the habitat and locality, a particular species of caterpillar may associate with a variety of ant species.

Our observations supported the idea that caterpillar calls attract ants in general, not one species in particular. Once the ants are in attendance, the calls act along with other ant organs to maintain a constant guard of ants. In the *Maculinea-Myrmica* systems, chemical cues are probably most important for governing the species-specific symbioses.

Because the symbioses between caterpillars and ants are so remarkable, I became curious about how these structures and the phenomenon of myrmecophily evolved in caterpillars. Among the thousands of butterflies and moths in the order Lepidoptera, the ability of caterpillars to form symbioses with ants is unknown outside the lycaenoids. The Riodinidae and Lycaenidae families show many parallels in their biology. Myrmecophilous caterpillars in both groups produce food secretions, chemical signals and acoustic calls. Conversely, nonmyrmecophilous species lack all or some of these behaviors and the associated ant organs. They are always mute, for example, even if they are closely related to calling, myrmecophilous species.

A comparison of the riodinid and lycaenid caterpillars with each other and of both types with other Lepidoptera suggests an evolutionary history that differs greatly from what Hinton envisioned. Although riodinid and lycaenid caterpillars have ant organs that secrete food and chemical signals, these organs develop on different body segments in each group. Lycaenid caterpillars produce calls, but they do not have vibratory papillae; the ant organs with which they call are still unknown. The ant organs of the riodinids and the lycaenids are therefore analogous but not homologous—they are similar but not evolutionarily related to each other. Comparative morphology alone thus suggests that myrmecophily evolved at least twice in butterflies: once in the lycaenids and once in the riodinids.

Moreover, the specific mechanisms mediating the symbioses may have evolved at least three times. Caterpillars of the riodinid genus *Eurybia* make calls and form symbioses with ants, but like lycaenids, they do not possess vibratory papillae. That fact implies that the ability of butterflies to produce calls must have evolved at least twice among the riodinid caterpillars alone.

The nonmyrmecophilous species are also instructive about the evolution of caterpillar-ant symbioses. The vast majority of butterfly and moth caterpillars bristle with spines, hairs or special behaviors that keep ants away. Lycaenoid caterpillars that do not associate with ants also have long hairs and lack ant organs. Overall, the species of riodinids and lycaenids that do not establish symbioses with ants are more similar to most Lepidoptera than to their own myrmecophilous cousins. This fact, too, suggests that myrmecophily evolved in lycaenoid butterflies at least twice and that the trait appeared relatively recently.

How a Caterpillar Manipulates Ants

Myrmecophilous caterpillars exploit the social and symbiotic behaviors of ants for their own purposes. Each of the specialized ant organs on the caterpillar promotes a symbiosis with ants in a particular way, either by mimicking the signals with which ants communicate or by offering fluids as symbiotic plants do. A look at the ants helps to explain some aspects of the development of the symbioses. Despite their apparent outward similarity, ant species are finely tuned specialists. The tendency toward specialization among ants is so well established that species are often divided into four general categories by their diets: predators, scavengers, seed eaters and other herbivores, and harvesters of secretions.

Most studies on the evolution of myrmecophily have considered all ant species as exclusively either predators of caterpillars or potential mutualists with them. To the contrary, however, my observations showed that relatively few of the more than 100 ant species at my study site in Panama either cared for or preyed on caterpillars. When I offered riodinid and lycaenid caterpillars to an assortment of ants, a few species killed the caterpillars and a few tended them, but most simply ignored them.

What unified all the ant species that tended caterpillars was their feeding ecology: much of their time was spent actively harvesting the secretions produced by insects in the order Homoptera and by plant extrafloral nectaries, in addition to those of butterfly caterpillars. Less than 10 percent of all the ant genera in the world associates with butterfly caterpillars, and this minority of ants also tends other secretion-producing insects and plants with extrafloral nectaries. Ant species that are general predators, specialist predators of arthropods, or herbivores and seed eaters are conspicuously absent from the list.

The observed patterns suggest that historically only a small fraction of the total ant diversity may ever have been involved in the evolution of symbioses with caterpillars. In all probability, these few ants favorably influenced the evolution of insects and plants that had the ability to produce secretions. Once secretions became important in the diet of particular ants, any other insect or plant capable of producing secretions could have easily garnered ants as bodyguards, thereby promoting the evolution of new symbioses with ants.

The dynamics of the symbioses between butterfly caterpillars, other secretion-producing insects, plants and certain ants bring to mind a further evolutionary concept. Because myrmecophilous insect herbivores and plants use the same ants as mobile defenses, they may be competing for the ants' attention. For example, when *Thisbe* caterpillars draft the ants on *Croton* plants as guards, they not only succeed in gaining their own protection, they also circumvent the defense of the plant: *Thisbe* caterpillars can feed on *Croton*, whereas most other herbivorous insects cannot. The caterpillars have inserted themselves into the ant-plant symbiosis and subtly undermined it. It is possible that the evolution of myrmecophily has permitted herbivorous insects to invade and exploit symbioses between plants and ants.

The study of interacting species provides new perspectives on the natural world and forces us to look at symbiotic associations in a much more dynamic context. As typically happens in ecological studies, a knowledge of other plant and insect species soon became essential for even a rudimentary understanding of the *Thisbe*-ant system.

Suggested Reading

DeVries, P. J. "Enhancement of Symbioses between Butterfly Caterpillars and Ants by Vibrational Communication," *Science* 248, no. 4959 (June 1, 1990): 1104–1106.

DeVries, P. J. "The Larval Ant-Organs of *Thisbe Irenea* (Lepidoptera: Riodinidae) and Their Effects upon Attending Ants," *Zoological Journal of the Linnean Society* 94, no. 4 (December 1988): 379–393.

DeVries, P. J. "The Mutualism between *Thisbe Irenea* Butterflies and Ants, and the Role of Ant Ecology in the Evolution of Larval-Ant Associations," *Biological Journal of the Linnean Society* 43, no. 3 (July 1991): 179–195.

DeVries, P. J., and I. Baker Butterfly. "Exploitation of an Ant-Plant Mutualism: Adding Insult to Herbivory," *Journal of the New York Entomological Society* 97, no. 3 (July 1989): 332–340.

Pierce, N. E. "The Evolution and Biogeography of Associations between Lycaenid Butterflies and Ants." *Oxford Surveys in Evolutionary Biology* 4. New York: Oxford University Press, 1987.

Thomas, J. A., G. W. Elmes, J. C. Wardlaw, and M. Woyciechowski. "Host Specificity among *Maculinea* Butterflies in *Myrmica* Ant Nests." *Oecologia* 79, no. 4 (June 1989): 452–457.

Chapter 5

Wings and Webs

Mating Strategies of Spiders

Ken Preston-Mafham and Rod Preston-Mafham

A spider's reproductive system bears little resemblance to that of any other group in the animal kingdom. From its head projects a pair of appendages called the pedipalps, sensory structures used primarily for tasting prospective prey. In male spiders the terminal segments of the "palps" are modified for introducing semen into a female.

The palpal organ, the basic copulatory structure of the male palp, can be

Offering the corpse of a rival, a common orb-weaver (*Meta segmentata*) male conducts the courtship ritual. Several males often fight for position in a female's web, leading to fatalities. This male has turned his victory to good account, using the loser's silk-wrapped body as a nuptial gift to the female. Normally he would wait for a fly to fall into the web. Range: Europe, temperate Asia and Canada.

likened to a simple bulb pipette. From a chamber that acts as a reservoir runs a narrow tube called the embolus with a pointed, open end through which semen passes. The male introduces the embolus into the female reproductive opening during mating. In higher families of spiders the palpal organ is surrounded by a complex set of sclerotized plates, hooks and spines. The females have simultaneously evolved a sclerotized structure, the epigyne, near the reproductive opening. The projections on the male palp will fit only into the epigyne of a female of the same species.

The palpal organs are not directly connected to the testes, so before mating can take place the male must somehow fill the palpal reservoirs with semen. First the male deposits a drop of semen onto silk from his spinnerets. This silk may be a single line stretched between a pair of legs, a simple, loose web held in a similar fashion or, in the higher spiders, a specially constructed web built onto an adjacent substrate. Then, holding the tip of the embolus against the semen, the male pumps it into the reservoir. In some species the male fills both palpal organs at the same time; in others, he fills only one, returning after a bout of copulation to prime the other. Once charged with semen, the male goes in search of a suitably receptive female. (In the family Linyphiidae, however, the males start sperm induction only when stimulated by contact with pheromones on the female web.)

Male web-building spiders would seem to face an insurmountable obstacle in finding mates—they have to search in three dimensions. They are, however, assisted by the females, which advertise the presence of their webs by releasing guidance pheromones. Female spiders that do not build permanent webs but wander around in search of prey always trail a single strand of silk dragline impregnated with pheromones. A wandering male of the same species coming across the dragline is thus able to follow it until he finds her. Females of the American fishing spider *Dolomedes triton* release their pheromones into the water near where they hunt, allowing male spiders to find them.

Spiders may also take part in some form of courtship. It can vary from the perfunctory, where the male just walks up to the female and induces her to accept him, to the prolonged, where the male may approach and be rebuffed many times before ultimately getting to mate. In large-eyed, active

Laying down lace of fine silk, a male giant wood spider (*Nephila maculata*) treads on the back of the huge female. Although she generally ignores him, the female sometimes tries to flick him off. When she does succeed, she begins to undo his painstaking handiwork. Perhaps to avoid such interruptions, males tend to lay silk while the female is busy feeding. The strands, which are applied near the female's scent receptors, are saturated with pheromones and may convey a sexual message. Range: South Asia to Japan and northern Australia.

Binding his mate to a leaf, a male European crab spider (*Xysticus cristatus*) will walk all over the much larger but cooperative female. He then climbs beneath her and mates for about an hour, after which he strolls off. The female is able to disentangle herself with relative ease, indicating that the tether is not a safety device for the male. Its purpose is unknown. Range: Europe, North Africa and much of Asia.

hunters, such as wolf and jumping spiders, the males signal their intentions to the females with their sometimes colorfully marked palps. Some males physically subdue females before mating.

Mating can certainly be dangerous for some males. In a limited number of species—the widows in particular—the female often eats the male after copulation. In many spiders, however, males and females have jaws of the same size. Because the male is considerably more agile, the female may be at as much risk of being attacked by him as he is by her. Moreover, some female spiders are many times the size of the males, yet the latter seem quite happy to reside in the same web.

All in all, these odd mating routines have proved extremely successful for the spiders.

Africanized Bees in the U.S.

Thomas E. Rinderer, Benjamin P. Oldroyd and Walter S. Sheppard

The long-anticipated announcement came in October 1990. Africanized honeybees, more popularly known as killer bees (because of sensationalized accounts of their attacks on people and animals), had finally crossed the Mexican border into the U.S. Less than 35 years after members of a honeybee subspecies living in Africa (*Apis mellifera scutellata*) were released outside São Paulo, Brazil, their descendants—the Africanized bees—had migrated as far north as southern Texas. They reached Arizona in 1993 and are expected to colonize parts of the southern U.S. before being stopped by climatic limits.

Their arrival in the U.S. raises many questions. How will the newcomers affect public health and the beekeeping industry? Why were African bees brought to the Americans in the first place? What allowed their progeny to be so extraordinarily successful? And, most important, can anything be done to minimize the impact of settlement by Africanized bees in the U.S.? We and others have devoted a great deal of study to this last question. That work, particularly research exploring the genetic makeup of the insects heading for the U.S., offers hope that efforts to control mating between Africanized honeybees and honeybees common in North American apiaries can be of considerable value.

One already obvious effect of the bees' arrival is heightened concern for public safety. Africanized bees typically defend their hives much more vigorously than do honeybee strains in North America. North American honeybees descend from rather gentle subspecies of A. mellifera that were imported primarily from Europe, when early settlers found that the New World lacked native honeybees. Compared with the European bees, those with markedly African traits become aroused more readily and are more prone to sting any person or animal they perceive is threatening their nest.

They may also attack in larger numbers (occasionally by the thousands) and persist in the attack longer (sometimes for hours).

Such behavior reportedly caused one human death in the U.S. and perhaps 1,000 in the western hemisphere, and it is responsible for many more fatalities among domesticated animals. Fortunately, most nonallergic individuals will survive an attack if they can run away and so limit the number of stings they suffer. Almost all individuals killed by Africanized honeybees have died because they could not flee—either because they had fallen and injured themselves or had otherwise become trapped.

Beyond posing a public health problem, the bees also promise to threaten the livelihood of thousands of commercial beekeepers (apiculturists) and farmers. Amateur and professional beekeepers alike keep their hives outdoors. It is therefore possible that European queen bees will mate with Africanized drones (males) and thereby introduce increased levels of defensiveness and other costly and troublesome traits into apiary colonies.

The queen's mating choices account for the characteristics of a colony because it is she who lays the eggs. Early in life, she mates in flight with perhaps 15 drones from other colonies and then never mates again. When bees are needed in a colony, the queen lays eggs into individual cells. Fertilized eggs usually give rise to worker bees—females that carry chromosomes from each of their parents and are responsible for foraging and guarding the nest. (If the larva emerging from a fertilized egg is fed a special diet, however, it can develop into a queen.) Unfertilized eggs give rise to drones; these males bear a single set of chromosomes (from the mother), and they die after mating.

If beekeepers are unable to control the infusion of undesirable traits produced by mating between European queens and Africanized drones, their profits will shrink, partly because measures will have to be adopted to protect workers and the public from excessive stinging. For instance, apiaries might have to relocate to sparsely populated areas, and everyone handling the bees will have to wear sturdy protective gear. Moreover, the bees tend to abandon hives more readily than do European bees; repopulating hives can be expensive.

Beekeepers could also face a reduction in honey production, which now amounts to about 200 million pounds annually (representing roughly $100 million in sales). Much research suggests that under climatic and ecological conditions that foster the abundant production of honey by European bees, Africanized bees would be less productive.

Meanwhile beekeepers who rent their colonies to farmers for the pollination of such crops as almonds, blueberries, apples and cucumbers would face additional financial losses. Rentals generate an estimated $40 million in fees every year, much of which goes to migratory beekeepers, who truck thousands of colonies to distant sites. Beyond having to exercise particular care to protect the public, beekeepers who maintained many Africanized bees could be prevented from bringing bees into non-Africanized areas.

Today's concerns are an outgrowth of an unfortunate series of events that began in the mid-1950s, after the government of Brazil decided to shore up that nation's beekeeping industry. At the time, European honeybees formed the basis for a strong beekeeping industry in many places, but not in Brazil. Brazil had only a small apiculture industry, partly because European honeybees were poorly suited to the country's tropical climate. Rarely, if ever, did a colony survive in the wild, and only considerable effort enabled beekeepers to sustain colonies throughout the year.

It is now evident that the poor performance by the European bees was related to their misreading of environmental cues for reproduction. Direct and indirect studies of genetics indicate that European bees, like all subspecies of *A. mellifera*, trace their ancestry to an Asian species that evolved the ability to regulate body temperature and survive in a temperate climate. The bees withstood the cold mainly by clustering together in sheltered nests and eating stores of honey they had collected in warmer seasons. Later they expanded their range to include Asia Minor, Europe and Africa, ultimately forming 20 or more subspecies adapted to particular locales.

In the course of evolution the behavior of the various European subspecies apparently became highly linked to seasonal fluctuations in day length. When hours of daylight begin to increase, heralding the imminent appearance of flowers, European honeybee colonies expand the size of their worker populations. By the time the flowers bloom, many workers are available to forage for pollen and nectar. Nectar, which contains a great deal of sugar, is converted to honey—a prime source of energy.

Linkage of the life cycle to day length works well in temperate regions, but in Brazil day length bears little relation to the availability of pollen and nectar. The rainy periods that are required for abundant production of flowers do not necessarily coincide with periods of extended daylight. Consequently, European colonies can be induced to expand even when food supplies are too scarce to support large populations.

In 1956 the best solution to Brazil's beekeeping woes seemed to be importation of a honeybee variant more accustomed to tropical living. The government therefore authorized Warwick E. Kerr, then at the University of São Paulo at Piracicaba, to bring *A. m. scutellata* from the highlands of eastern and southern Africa for study. Kerr obtained 170 queens, although only 46 from South Africa and one from Tanzania survived the journey from South Africa to a research apiary in Rio Claro. (Rio Claro lies roughly 100 miles from São Paulo.) He chose individuals that had already mated with African drones and were thus ready to lay the eggs needed to create complete colonies.

In 1957, within months after the African colonies were established, a visitor to the experimental apiary removed screens that had been placed at hive entrances to block queens from leaving. The reasons for the removal are unclear, but before the act was discovered, 26 colonies had abandoned their hives with their queens. For years, those liberated colonies were thought to have been the founders of the entire Africanized population. However, scientists have learned that soon after the initial release, queens reared from the remaining colonies were distributed to beekeepers in Brazil. The additional releases undoubtedly helped to ensure that enough African insects would be available to establish permanent feral populations of Africanized honeybees in Brazil.

The freed bees and their descendants found Brazil to be a hospitable place, and so they thrived. Compared with European bees, the newer arrivals were better able to take their reproductive cues from variations in the availability of rainfall and flowers and were better equipped to cope with dry seasons. When flowers are abundant, Africanized colonies engage in a process known as reproductive swarming: the queen and a good many hive members split off to establish a new, growing colony. This swarm leaves the remainder of the original colony with a young queen, who repopulates the hive. When floral resources dwindle severely, Africanized bees are likely to abscond—they gather any remaining honey and abandon the hive en masse, to try to find a more hospitable locale. (European bees, in contrast, swarm perhaps once a year and rarely abscond.)

As the Africanized bees flourished in Brazil, they fanned out in all directions, including into areas that previously had no beekeeping. In the 1960s they began to draw an increasing amount of attention, especially for their intense nest defense, and it became clear that they could be very troublesome.

By 1972 the U.S. government began considering the potential impact of the bees on the U.S.

In the mid- to late 1980s the U.S. government, with the cooperation of Mexico, decided to try retarding the spread of the bees into the U.S. by establishing a "bee-regulated zone." The final plan called for detecting and killing any swarms that arrived in parts of Mexico the bees would have to traverse in order to reach the U.S. Combined with weather inhospitable to migration, that effort (which proved more difficult to implement than had been hoped) may well have delayed the arrival of the bees for a while. But it was clear they were not going to be stopped altogether.

Interestingly, as anxiety mounted in the U.S., Brazilians found a way to use Africanized bees for the intended purpose: to strengthen their beekeeping industry. Initially many beekeepers abandoned the craft. But the Brazilian government embarked on a campaign to teach potential beekeepers how to cope and to instruct the public about how to avoid the bees and handle attacks. A new generation of apiculturists emerged. Indeed, in some parts of Brazil that were once unable to sustain European honeybees, people earn their livelihood through keeping Africanized bees and harvesting their honey. These individuals maintain reasonable traits in their stocks by destroying queens in the most defensive and least productive colonies.

Unlike Brazil of the 1950s, the U.S. has little to gain from settlement by Africanized bees. In theory, two major strategies might be helpful. Certainly, beekeepers could protect their stocks to some extent by practicing "requeening" frequently. The procedure involves inducing colonies to accept substitute queens of a beekeeper's choosing, often purchased from breeders of queens. Many apiculturists are already adept at requeening. They use it to increase the production of offspring (replacing old, less productive queens with new ones) or to control the genetic makeup, and hence the characteristics, of hive populations.

Another protective strategy, known as drone flooding, calls for maintaining large numbers of European drones in areas where commercially reared queen bees are mated. Even if the areas have been invaded by Africanized émigrés, the vast number of European males would ensure that European queens would mate almost entirely with European drones.

Of course, the drone-flooding strategy assumes that honeybees bearing essentially African genes and those bearing essentially European genes can hybridize—that is, mate with each other and produce viable offspring bearing genes and traits from both parents. But can the two groups in fact hybridize?

Morphological comparisons are much more difficult than they might sound. Even if one examines the extremes—African bees living in Africa and European bees living in Europe—the two groups look alike. But colonies can be distinguished by a statistical procedure called multivariate discriminant analysis. In doing such an analysis, researchers measure many different body parts—among them, the length and width of the wings and leg segments, and the angles at which various veins intersect in the wings. Although the mean scores for African and European samples will not differ significantly on any one measure, combining group means for many measures makes it possible to distinguish overall differences that do exist.

To assess whether the invading bees in Central and South America differed physically from A. m. scutellata—which would suggest genetic mixing had taken place—we compared their final scores with those attained for African and European bees. The comparison revealed that feral populations in Mexico, Brazil, Argentina and Venezuela resembled both European and African bees to various degrees, although they were more like the African bees.

Similarly, when we plotted summed measures for three different clusters of traits against one another on three axes, the point representing purely African bees fell far from that representing purely European bees in North America. The points representing bees from Mexico, Brazil, Argentina and Venezuela fell in between, roughly a third of the way between those two extremes but closer to the African value. These findings suggested that the populations advancing toward the U.S. were not pure Old World African stock; they were indeed hybrids that had acquired some European genes in their travels.

Yet the findings were contradicted by other observations. Notably, bees in the colonized areas seemed to display clearly African traits, namely, intense defensiveness and frequent swarming and absconding. If hybridization was going on, it certainly was not obvious behaviorally. The first direct genetic studies raised similar doubts. They compared mitochondrial DNA in bees from colonized areas with that in European and African bees. Mitochondria, the energy factories of cells, contain small rings of DNA that are distinct from the chromosomal DNA housed in the nucleus. Nuclear DNA directs the emergence of physical and behavioral traits. Mitochondrial DNA provides about a dozen genes required strictly for energy production. Most animal species (including humans) inherit their mitochondria, and thus mitochondrial DNA, exclusively from the mother.

It turns out that African and European bees differ slightly in the sequence of nucleotides (the building blocks of DNA) in their mitochondrial DNA. Hence, by identifying known markers of the variable DNA segments from bees in Africanized areas, investigators were able to trace the maternal lineage of the insects to either Africa or Europe. (The markers used are DNA fragments that form when mitochondrial DNA is cleaved by a restriction enzyme. For example, one fragment generated from African DNA appears as two smaller fragments in European DNA.)

Determining whether hybridization could in fact occur required investigation of bee colonies from areas known to have been supporting European honeybees when the newcomers arrived. We therefore traveled to Argentina, which lies west of the thin, southernmost part of Brazil and extends much farther south, into a temperate zone. Africanized bees have not become established in the southern half of the country, which supports abundant beekeeping with European strains. But they have established large populations in the northern half, particularly in the topmost quarter of the country, which has never maintained as many European bees.

Julio A. Mazzoli, a graduate student from the University of Buenos Aires, helped us find more than 100 colonies in bridges, trees, electric utility poles, fruit boxes and other enclosed areas favored by honeybees. The collection included representatives from areas extending from the north into the south. Back in our laboratories, we evaluated the morphology and the composition of mitochondrial DNA. Our collective results showed decisively that hybridization had taken place. As part of our evidence, we found that a large number of the sampled colonies had physical features intermediate between those of European and African bees. Further, more than a quarter of the colonies either displayed African morphology (reflecting the activity of nuclear genes derived from African ancestors) yet bore European mitochondrial DNA (reflecting the influence of a female European ancestor) or else displayed European morphology yet bore African mitochondrial DNA.

The morphological work yielded another interesting finding: European physical features were more prominent in the southern, temperate regions of the studied territories than in the northern corner, where African morphology predominated. But in a band of fairly temperate territory between those areas, no single cluster of morphological features predominated. This mixture of traits implied that hybridization had occurred extensively in the intervening zone, a conclusion supported by isozyme studies. We found relatively few hybrids outside the transition zone presumably because conditions in the tropical north favor survival of bees having primarily African traits, whereas conditions in the temperate south favor survival of bees having primarily European traits.

Such selective pressures may lead to a similar pattern in the U.S., where the southernmost regions have a subtropical climate and northern areas are temperate. European-like bees may be less competitive in the Deep South, and African-like bees should be less competitive in the North. In the intervening central regions, there may be a mixture of hybrids whose gentleness and tolerance of cold increase with increasing latitude. It is also possible that hybrids will be abundant in some central areas during the warm seasons but will disappear in the winter.

Because there were few European bees in the tropical regions of Argentina, we could not determine whether the presence of a sufficiently large European population would cause African-like bees to mate with them and produce hybrids that survived and reproduced well in the tropics. We knew only that such behavior was commonplace in temperate areas.

We sought an answer in the Yucatán Peninsula of Mexico. The peninsula has an ideal combination of a tropical environment and an extensive, long-established population of European honeybees. In fact, the Yucatán has the greatest concentration of commercial honeybee colonies in the world. This was the first massive population of European bees encountered by the expanding populations of Africanized bees as they migrated north from Brazil.

We again collected samples from a large tract. This time we relied on the cooperation of beekeepers, who own most of the bees in the Yucatán. Despite being cared for by humans, the honeybees in the Yucatán are probably the genetic equal of feral bees. Beekeepers obtain them by putting out boxes the insects can colonize. Owners usually make little effort to control the genetic makeup of the hive.

All but a few of our samples came from colonies that had not undergone requeening in the two years since Africanized bees had first been detected on the peninsula. Most of the insects still possessed clearly European morphology, but some possessed mainly African morphology, and many had intermediate morphologies indicative of hybridization. Mitochondrial analyses provided still more evidence of interbreeding: a number of colonies displayed either European morphology and African mitochondrial DNA, or the reverse. Thus, a tropical environment does not appear to pose an unbreachable barrier to hybridization.

The least evidence of African traits appeared in the few colonies that had been requeened. This simple observation implies that requeening—one of the chief tools beekeepers have for controlling the Africanization of their stocks—can certainly be helpful.

The potential of Africanized honeybees to hybridize successfully with

European honeybees is good news for beekeeping. We anticipate that frequent requeening of commercial colonies and drone flooding in commercial queen-breeding areas would serve to dampen the acquisition of unwanted African traits. We should note, though, that there are dissenters who contend that hybridization efforts will fail to prevent the eventual widespread introduction of dramatic African traits into honeybee populations.

If we are correct that Africanization of U.S. apiaries can be limited, then it seems that, with care, the practice of transporting bees to crops could be continued safely without leading to the significant establishment of Africanized bee colonies in new territories. Fortunately, there are ways to assess the character of individual colonies, and these methods could be employed to guarantee that colonies moved from place to place are European.

It is inevitable that the incursion of Africanized bees into the U.S. will increase the costs of managing commercial colonies, at least temporarily. It is also likely that some African genes will spread through feral and managed bee colonies. Yet vigilance and coordination by apiculturists have every chance of preserving the European behavior of commercial honeybee stocks, thereby reducing the damaging effects of Africanized insects on beekeeping and allaying the fears of the public.

Predatory Wasps

Howard E. Evans

The sting of bees, wasps and certain ants gives these insects a formidable defense against predators. In the solitary wasps, however, the sting still serves primarily the function for which it originally evolved: the taking of prey. The solitary wasps are hunters. Most familiar in the "digger" or "mud dauber" forms, they have diversified into hundreds of genera and thousands of species in pursuit of their fellow insects and other arthropods, such as the spiders. Each species tends to specialize, hunting down a particular prey and disregarding others of similar size and more ready availability. Some restrict their predation to a single species or genus, and only a few claim victims from more than one order of arthropods. The affinity of predator to prey is, in each case, as characteristic of the wasp as the anatomical features that distinguish it from other wasps. Such specialization of behavior is not surprising in view of the difference in strategy and tactics required for capturing a caterpillar, for example, compared with a fly. The consequent diversity in the behavioral repertories of the solitary wasps has invited increasing attention in recent years, as zoologists have turned to the study of the decisive role of behavior in the origin of species.

Solitary wasps are predators of a rather special sort. Only a few take prey as food for themselves; for the most part the adults of all species feed on sugars in solution, which they find in the nectar of flowers, in ripe fruit or in the honeydew secreted by aphids and other plant-sucking insects. The male wasps, in fact, are not predators at all and feed exclusively on plant exudates. It is the females that take prey, and they do so primarily to feed their larvae. In this remarkable plan of behavior the solitary wasps foreshadow the still more elaborate larvae-nurturing of the social Hymenoptera—the ants, bees and social wasps, all of which apparently arose from certain long extinct groups of solitary wasps. (The termites arose from an entirely different stock, the cockroaches of the order Orthoptera.) The solitary wasps in turn have derived the elements of their behavior—their specificity as to prey, the restriction of the predatory habit to the female and the consignment of the prey to the nurture of the next generation—from their precursors in the Hymenoptera line.

The most primitive Hymenoptera, on the basis of many features of larval and adult structure, are the sawflies, and these first appear in the fossil record at the beginning of the Mesozoic era, some 200 million years ago. With its sawlike ovipositor the female deposits its eggs in plant tissue. Most species are highly host-specific. In the Jurassic, the second period of the Mesozoic, the parasitic Hymenoptera, including the ichneumon wasps, made their appearance. The female deposits its eggs on or in another arthropod (most commonly a plant feeder); the larvae feed on the host, causing its death only when they have completed their own development. These insects flourished almost immediately and even today form an enormous group of many thousands of species. Some are equipped with ovipositors twice the length of their bodies, with which they are able to plant their eggs on grubs burrowing deep in the trunks of trees.

For this unique form of parasitism O. M. Reuter coined the term "parasitoid." When the true wasps appeared toward the end of the Mesozoic, they inherited this manner of life; some living groups of primitive wasps still behave essentially like parasitoids. But most wasps paralyze their hosts (now called "prey") by stinging them, then store them in the nest where the egg is laid and larval development takes place. The true wasps are no more predators in the usual sense than the ichneumon wasps are parasites. One might perhaps use the term "predatoid" to epitomize the origins and gross behavior of these insects.

The adaptation of a particular wasp to its prey presents one of the most intriguing problems in the study of behavior. In two widely separated localities, the digger wasp *Aphilanthops laticinctus* has been found to prey exclusively on one species: the prairie mound-building ant *Pogonomyrmex occidentalis*. A related species, *A. haigi*, seems to confine its predation to another ant, *P. barbatus rugosus*. I have seen *haigi* hunting in areas where a closely related ant, *P. maricopa*, was more abundant, and I have seen these wasps approach a worker *maricopa*, back away and proceed to hunt *barbatus rugosus* workers. *Maricopa* appears, however, to be the normal prey of another species, called *A. sculleni*.

One *Aphilanthops*, *A. frigidus*, a common species in the eastern U.S., is known to prey on queen ants instead of workers; in museums there are many specimens pinned with their prey. Curiously, this wasp takes only winged queens at the time of their nuptial flight and ignores the queens that have lost their wings and are seen at times running in numbers over the ground. The wasp removes the wings from the ant, however, before placing it in the brood cell. Without exception the victims belong to the species *Formica fusca* or one of two or three other closely related species of the genus *Formica*.

Wasps more commonly prey on several related species of a genus— *Cerceris halone*, for example, preys on the long-beaked weevils of the genus *Curculio*. Still more commonly, a given species of wasp will prey on several or even many genera of one family or several related families. The great golden digger, *Sphex ichneumoneus*, takes katydids and long-horned grasshoppers of several genera in at least two subfamilies. Less commonly, a wasp may prey on a wide variety of insects of a given order. *Philanthus pacificus* has been found to take prey belonging to six different families of Hymenoptera, including bees, ichneumonids and even other wasps.

Such host-specificity is more curious because the larvae of wasps can often be induced to develop normally on prey quite different from that provided by the mother. Jean Henri Fabre successfully reared *Bembix*, a predator of flies, on a diet of grasshoppers. Similar experiments have been performed by Fabre and other investigators on other species of wasps. The host must, of course, be paralyzed or freshly killed if it is to be acceptable.

For a wasp to take insects of more than one order is unusual, but several species are known to do so. *Crossocerus quadrimaculatus* preys on a wide variety of flies, moths and caddis flies. *Lindenius columbianus errans* fills its nest with a remarkable mixture of midges, small parasitic wasps and minute bugs. Another unusual wasp is *Microbembex*, which is known to stock its brood cells with insects of 10 different orders as well as with spiders. The victims in this case, however, are picked up from the sand already dead or disabled; *Microbembex* is the only genus of digger wasps that has become a scavenger rather than a predator.

Many species of wasps take their prey in habitats quite different from and often distant from that in which the new adult emerges from the brood cell and where she eventually builds the brood cells of her offspring. Little is known about her initial hunting flights. These must often involve much random flying about before the wasp locates the source of prey appropriate to her species. Once she arrives in the proper habitat it is apparent that an assortment of cues direct her to her prey. In the case of the bee wolf, *Philanthus triangulum*, the wasp flies from one flower to another; when she sees a moving object on a flower about the size of a bee, she hovers a short distance downwind from it. At this point olfactory cues become important: if the object has the odor of a bee, the wasp pounces on it. But she does not actually sting it unless she finds it has the "feel" of a bee. Less thorough studies of a number of other digger wasps suggest that this succession of visual-olfactory-tactile cues may serve to guide wasps of many different species to their prey. The tendency to respond to appropriate cues in each of these sensory modalities is apparently part of the unlearned, genetically determined behavioral repertory of each species. William Morton Wheeler University many years ago suggested that the olfactory system of the wasp is conditioned in the larval stage, when it consumes the specimens of its future prey captured for it by its mother. This seems improbable in view of what has already been said about the diversity of prey taken by some species; in any case, it suggests no corresponding way in which the visual components in this behavior might be "conditioned."

Wasps do not normally "make mistakes" with respect to prey. I have accumulated thousands of prey records of various North American species of *Bembix*, and not a single one involves an insect other than a fly (Diptera). Yet in different parts of its range any one species of *Bembix* will hunt in quite different habitats and take quite a different array of flies.

Although one is most likely to see wasps hunting in broad daylight in the open air, this must not obscure the fact that important species do their hunting under entirely different circumstances. Wasps that prey on crickets and cockroaches do most of their hunting on the ground, creeping into crevices under stones and fallen timber and into holes in the ground. Diggers of the genus *Podalonia* unearth cutworms from the soil. One species of *Bembix* does all its hunting just at twilight, preying on Diptera that have come to rest for the night in vegetation.

The stinging of the prey that follows capture is a relatively stereotyped sequence of motions that differs from one major group of wasps to another in apparent adaptation to the anatomy and physiology of the prey. *Liris nigra* stings its cricket prey first in the vicinity of the nerve ganglion controlling the hind legs—the jumping legs—and then twice more in the vicinity of the ganglia controlling the other two pairs of legs. The hunters of caterpillars sting their prey not only on the thorax but also several times along the underside of the abdomen. In caterpillars the "prolegs" on the abdominal segments of the body are as important in locomotion as the thoracic legs, if not more so, and the nervous system is not highly concentrated in the thorax.

The extensive literature that explains the stinging behavior of wasps in terms of the neural anatomy of the victim may have to undergo revision as the result of the work of Werner Rathmayer. He observed that the bee wolf inserts its sting into the honeybee only once, in the membrane around the coxae, or anchor segments, of the front legs on the underside of the thorax, and that the stinging lasts from 20 to 50 seconds. By careful dissection of the victims Rathmayer was able to show that the sting does not normally penetrate the ganglion. Instead the venom diffuses from the point of injection into the flight muscles and the muscles controlling the legs.

Rathmayer's finding that the venom does not act directly on or through the nervous system is well documented. It may be that the stings are inflicted through the underside of the thorax not because of the presence there of nerve ganglia but because "chinks in the armor" around the coxae give ready access to the muscles of locomotion.

In the task of classifying wasps, the study of behavior and particularly of prey selection has shed light on difficult problems. Workers in our laboratory divided one genus of the tribe Bembicini into two: *Stictiella*, which has been found to provision its nests exclusively with moths and butterflies, and *Glenostictia*, which feeds its larvae from day to day on flies.

The major steps in the evolution of wasps were taken millions of years ago. Many of the wasps found in Baltic amber look much like contemporary wasps, and it is probably safe to assume that many of them behaved much the way their relatives do today. Hence it is not surprising to find that those groups of wasps that are ranked as primitive on the basis of their structure and generally simpler behavior patterns prey largely on orders of insects considered low on the evolutionary scale. The Ampulicinae, for example, are an isolated and probably relict group of digger wasps; they limit their predation to cockroaches, dominant insects of earlier geologic time. Other roach-hunters turn up in the relatively primitive subfamilies Sphecinae and Larrinae, most of whose members are adapted to the now more dominant groups of Orthoptera, such as crickets and grasshoppers. Spiders, an arthropod order of ancient origin, also serve as prey for certain Sphecinae and Larrinae and do not attract predation by more advanced digger wasps. Correspondingly, the more advanced wasps take as their prey more advanced insects, such as flies, bees and beetles, that go through complete metamorphosis. The higher flies underwent most of their radiation during the Tertiary period, which began some 60 million years ago, and several groups of structurally advanced wasps quickly took advantage of them.

In the sequence of evolutionary development that begins to emerge from study of these relations, the size of the prey compared with the size of the wasp emerges as a significant factor. Once the prey has been immobilized it must be carried back to the nest. Obviously the size of the prey must strongly influence the mode in which the wasp transports it. Since primitive wasps generally install only a single victim in each brood cell, they must take prey as large as or larger than themselves if their larvae are to have enough food to reach full size. The more primitive wasps accordingly drag their prey over the ground, grasping the victim in their mandibles. Some beat their wings to facilitate their progress, and some drag their prey up on a high object and glide off with it. As might be expected, they cannot cover much ground by these methods, and they nest correspondingly close to the habitat of their prey. Wasps of the genus *Priononyx*, for example, nest in bare spots on the prairie, where their grasshopper prey abounds.

Wasps of the related genus *Sphex* illustrate the first stage of progress beyond this simple pattern of behavior. They hunt green meadow grasshoppers in tall weeds and bushes, and yet they make their nests in bare sand and gravel. They are able to transport their victims considerable distances in flight because they take smaller prey. And because the grasshoppers are smaller the wasps must supply several of them to each larva. A species of the genus *Pemphredon* similarly may gather aphids in habitats far removed from the soft wood in which it constructs its galleries; this wasp flies back and forth between its hunting grounds and nest many times a day.

All of these species carry their prey in their mandibles. As a result they cannot dig while holding their prey, because it is held so far forward that it obstructs the use of their front legs. These wasps must leave their nests open or, if they close them, must put down the prey while they scrape open the entrance. Either of those actions exposes the prey or the contents of the nest to the attacks of a variety of parasites. The next breakthrough in the evolution of the digger wasps came, therefore, with the development of what I have called pedal prey-carrying mechanisms. Wasps that have made this advance grasp their prey in their middle and hind legs or in their middle or hind legs only, thereby leaving their front legs and mandibles free. They close the nest entrance when they leave, and they are able to open it readily when they return while still holding their prey. Since the prey is carried well back beneath the body close to the center of equilibrium in flight, the pedal mechanisms greatly enhance the carrying efficiency of these wasps. Four subfamilies have made this breakthrough, and some members of two of those subfamilies have made the further advance to what I have called abdominal mechanisms, carrying the prey on their sting and, in one genus, on special modifications of the segments at the end of their abdomen. This significant trend in the evolution of more advanced wasps would not have been possible without a parallel adaptation toward smaller prey.

A listing of the major groups of digger wasps in the probable order of their emergence as indicated by their morphology shows excellent concordance with changes in type of prey and in hunting and prey-carrying behavior. The crossing over of a line of wasps to a new type of prey clearly represents the invasion and occupation of a new adaptive zone.

What is remarkable about the solitary wasps is the thoroughness with which they pursued the evolutionary diversification of the arthropod phylum. Many of the arthropods preyed on are themselves predators; others exhibit protective coloration that appears to protect them from predators other than wasps; still others are mimics of stinging insects or are armed with chemical defense mechanisms that seem to hold other predators at bay. Diptera-hunters such as *Bembix* capture drone flies and other bee mimics, but bee-hunters such as *Philanthus* spurn bee mimics even under experimental conditions. Many formidable stingers, including worker harvester ants, bees of many kinds (not excluding bumblebees) and other wasps (not excluding social forms), are among the prey of solitary wasps. Evidently the various mechanisms of defense, deception and concealment elaborated by the insect world are primarily adapted to protect their possessors from predation by vertebrates, since they do not protect them from the solitary wasps.

The solitary wasps are esteemed by entomologists not only as subjects but also as colleagues of a sort. A small species of *Philanthus* nesting in my backyard a few years ago collected three new species of bees; these have since been described, one of them being named appropriately *philanthanus*. The prize example of this collaboration comes from the Congo. In 1915 Herbert Lang trained several native assistants to collect flies being brought to their nests by *Bembix dira*. Among the nearly 1,000 flies brought in by the wasps there were more than 200 species belonging to 14 families. A great many thus collected were new to science, and several have not been rediscovered since!

Suggested Reading

Fleming, Jo-Ann G. W. "Polydnaviruses: Mutualists and Pathogens," *Annual Review of Entomology* 37 (1992): 401–426.

Lavien, M. D., and N. E. Beckage. "Polydnaviruses: Potent Mediators of Host Insect Immune Dysfunction," *Parasitology Today* 11, no. 10 (1995): 368–378.

Quicke, Donald L. J. *Parasitic Wasps*. City: Chapman & Hall, 1997.

Rinderer, Thomas E. J., Anthony Stelzer, Benjamin P. Oldroyd, Steven M. Buco, and William L. Rubink. "Hybridization between European and Africanized Honey Bees in the Neo-tropical Yucatan Peninsula," *Science* 253 (July 19, 1991): 309–311.

Spivak, Marla, David J. C. Fletcher, and Michael D. Breed, ed. 1991. *The "African" Honey Bee*. Boulder, Colorado: Westview Press.

Sheppard, Walter S., Thomas E. Rinderer, Julio A. Mazzoli, J. Anthony Stelzer, and Hachiro Shimanuki. "Gene Flow between African- and European-Derived Honey Bee Populations in Argentina," *Nature* 349 (February 28, 1991): 782–784.

Taylor, Orley R., Jr. "The Past and Possible Future Spread of Africanized Honeybees in the Americas," *Bee World* 58, no. 1 (1977): 19–30.

Tumlinson, James H., W. Joe Lewis, and Louise E. M. Vet. "How Parasitic Wasps Find Their Hosts," *Scientific American* 268, no. 3 (March 1993): 100–106.

6 Extreme Plants

Plants That Warm Themselves

Roger S. Seymour

In the spring of 1972 George A. Bartholomew, a leader in the study of animal physiology, invited a group of his students and coworkers from the University of California at Los Angeles to a dinner party. Among his guests was Daniel K. Odell, now of Sea World in Florida. En route to the affair, Dan noticed some striking flowers. They consisted of a rather phallic projection that was about nine inches long and partly enveloped by a leaflike structure. Intrigued, he picked one to show the other partygoers. When he handed the cutting to Kenneth A. Nagy and me, we were astonished to find it was warm. What is more, the flower grew hotter as the evening progressed, appearing to become warmer than the human body. As zoologists, we were dumbfounded. How could a mere plant heat itself more than the pinnacle of organic evolution—the warm-blooded animal?

From that moment on, I hunted for and analyzed hot plants whenever I could steal time from my research into animals. Among our discoveries is that some plants produce as much heat for their weight as birds and insects in flight, the greatest heat producers of all. And a few plants actually thermoregulate, almost as if they were birds or mammals: they not only generate warmth; they alter their heat production to keep their temperature surprisingly constant in fluctuating air temperatures.

We were not the first to realize that some plants give off heat. When we delved into the botanical literature, we learned that almost 200 years earlier, in 1778, the French naturalist Jean-Baptiste de Lamarck reported that the European arum lily, probably *Arum italicum*, became warm when it flowered. This plant is a member of the huge family Araceae, which includes *Philodendron*, the kind of plant Dan had plucked. It also includes jack-in-the-pulpit, skunk cabbage and many other familiar plants. In these

Only three plants have yet to be shown to regulate their temperature. Such control is exhibited by the flowering parts of *Philodendron selloum, Symplocarpus foetidus* (skunk cabbage) and *Nelumbo nucifera* (sacred lotus).

so-called aroids, or arum lilies, the flowering part is termed a spadix and is not a true flower; it is an "inflorescence," or clustering of small flowers (florets). The aroid spadix, which consists of hundreds of florets assembled on a common stalk, is partly enveloped by a large bract, or specialized leaf, known as a spathe. Dan's "flower"—from *P. selloum*—was therefore not technically a flower; it was an inflorescence.

Scientists had subsequently discovered that other species of this bizarre family heat up, and they had noted weak heat production by a few plants outside the aroids—by the flowers of the Amazon water lily and of the custard apple, by the inflorescences of a few palms, and by the male cones of certain cycads (fernlike plants that resemble palms). Some investigators, among them Bastiaan J. D. Meeuse of the University of Washington, had even uncovered clues to how the cells of various plants generate warmth.

For instance, they found that to make heat, aroids activate two biochemical pathways in mitochondria, which are often called the power plants of cells. These pathways are distinguished by their sensitivity to cyanide. The one that can be poisoned by the chemical is common to plants and animals; the one that is insensitive to cyanide occurs in heat-producing plants, certain other plants, fungi and some unicellular organisms. Both pathways typically use nutrients and oxygen to manufacture an energy-rich molecule called ATP (adenosine triphosphate), which can subsequently be broken apart to provide energy for cellular activities or to produce heat. It is unclear, however, whether aroid cells that warm up generally do so by first making ATP and then breaking it down or by simply liberating heat directly from the pathways without producing ATP as an intermediate.

A "Warm-Blooded" *Philodendron*

We started off looking at *P. selloum* from an entirely different vantage. Instead of examining individual cells or molecules, as most botanists had done, we studied the inflorescences as if they were whole animals. Bartholomew's laboratory at U.C.L.A. conducted comparative studies on heat production and body temperature regulation in animals, and so all the needed equipment and methods were at hand. Furthermore, many *P. selloum* plants were in flower right outside our laboratory window, giving us easy access to our subjects.

Our earliest experiments aimed to determine whether the inflorescence truly had become as hot as we suspected at the party. We impaled spadices with temperature probes and connected the probes to a machine in the laboratory that recorded temperature continuously. During the measurement period, the air outside averaged about 20 degrees Celsius (68 degrees Fahrenheit), and the spadix temperature remained about 20 degrees C higher, near 40 degrees C (104 degrees F). The inflorescence was, indeed, hotter than its environment and hotter, too, than a person.

We wanted to know more, such as the range of *P. selloum*'s heat-producing capabilities. Because we could not control the air temperature outdoors, we cut some specimens and put them into indoor cabinets where we could vary the temperature as we chose. Indoors, we could also examine the plant's rate of heat production by the simple expedient of measuring its rate of oxygen consumption. We felt confident in applying consumption of oxygen as a gauge because of the intimate connection between oxygen use and heat generation. In animals, every milliliter of oxygen consumed results in

about 20 joules of heat. Thus, the rate of oxygen use can be converted readily to the rate of heat production in watts (joules per second).

We examined the inflorescences at air temperatures ranging from below
freezing to uncomfortably hot for humans. At the coldest extremes, some of
the inflorescences could not heat up at all. But their temperatures soared to
as high as 38 degrees C (100 degrees F) when the environmental temperature was still a cool four degrees C (39 degrees F)—a 34 degree C difference
between the inflorescence and the air. The cuttings became hotter still as the
air temperature rose further, but the difference between them and their environment became less dramatic. The inflorescences peaked at 46 degrees C
(115 degrees F) when the interior of the cabinet was a tropical 39 degrees C
(102 degrees F). Further, the estimated rate of heat production decreased as
the temperature of the environment increased.

The plant was obviously adjusting heat production to maintain warmth
in cold weather and to prevent overheating in hot conditions. The conclusion was startling: these inflorescences did more than produce heat. Like
warm-blooded birds and mammals, they were thermoregulating.

Roger M. Knutson reported that the spadix of the eastern skunk cabbage, *Symplocarpus foetidus*, holds its temperature between 15 and 22
degrees C (59 to almost 72 degrees F) for at least two weeks during February
and March, when air temperatures are below freezing. (The plant reportedly
melts snow around it.) And Paul Schultze-Motel and I discovered that the
sacred lotus, *Nelumbo nucifera*, maintains its temperature near 32 degrees
C (almost 90 degrees F) for two to four days in the middle of its summer
flowering period, even when air temperatures drop to 10 degrees C (50
degrees F). In this case, the spongy, cone-shaped center of the flower, called
the receptacle, produces most of the heat. The sacred lotus belongs to a
completely different family from *Philodendron* and skunk cabbage, suggesting that thermoregulation evolved independently in aroids and in the lotus.

The Value of Thermoregulating

Aroids and certain other plants heat themselves to vaporize scents that
attract insects. Vaporization of attractants could partly explain heating in
thermoregulating plants but would not explain why heat production is

raised and lowered to keep temperature within a set range. We can suggest two reasons for thermoregulation having evolved in some plant species.

First, it may create a warm, stable environment for pollinators and thereby facilitate reproduction. Large insects that carry pollen from one flower to another typically require high body temperatures for flight, locomotion and social interactions, and they often expend a great deal of energy keeping warm. Those that visit thermogenic flowers would be provided with a fairly steady level of heat directly from the plant. They could eat, digest, mate and function in other ways without having to squander energy of their own to stay warm. Alternatively, the flower itself may require a constant temperature for proper development of its own reproductive structures or to protect sensitive parts from damage that might occur if heat production were uncontrolled.

Either hypothesis could explain why a plant evolved the ability to thermoregulate. Yet the interaction between *P. selloum* and pollinating insects does lend some credence to the idea that thermoregulation in this plant may have been adopted because it abetted pollination.

The inflorescence of *P. selloum* contains three types of florets. At the top are fertile males that produce pollen. At the base are females capable of producing fruit when they are fertilized. Separating the fertile males and females is a band of sterile males that provide nourishment to pollinating insects and also furnish most of the inflorescence's heat. Tantalizingly, the 18- to 24-hour period of temperature regulation in the inflorescence overlaps the period of female receptivity to pollination. During these hours, the spathe surrounding the spadix opens widely and gives pollen-bearing insects—mainly beetles—easy access to the warm, sterile florets and nourishment. Then the spadix cools, and the spathe closes around it, trapping some beetles inside. After about 12 hours, by which time the female florets are sure to have been pollinated, the flower warms up again, the spathe reopens partway, and the fertile male florets shed their pollen. The pollen sticks to escaping insects, which fly off to repeat the cycle. This sequence promotes cross-pollination and prevents self-pollination; it thereby increases genetic diversity, which favors reproductive success.

In common with *P. selloum*, the sacred lotus maintains high temperatures when the female stigmas are moist and receptive and before the pollen is shed. Heating begins before the petals open widely and ends when opening is complete. The shape of this flower is also appropriate for pollination by beetles. But whether thermoregulation evolved specifically to aid beetles in that endeavor is unclear. The uncertainty arises because we do not know the native habitat of this plant and whether beetles are the main pollinators in that area. Further, the flower clearly is not dependent on beetles; it can be pollinated by other insects after the petals open widely and heat production subsides.

How *Philodendron* Operates

The question of how plants thermoregulate is as fascinating as that of why they do it and was in fact my main concern when I began studying *Philodendron* seriously. The answer was not at all obvious, given that plants operate very differently from animals.

Plants have no fur, no feathers, no nervous system, no muscles, no lungs, no blood and no brain. How, then, we asked, does *P. selloum* raise and lower its temperature to keep the inflorescence in the 38 to 46 degree C range?

We first had to know which part of the inflorescence produced heat, a mystery at the time. We separated the three types of florets from their stalk and measured their oxygen consumption rates. On the basis of these measurements, we calculated the rate of heat production. The sterile males consumed the great bulk of the oxygen we supplied; the fertile males consumed a bit; and the females and the stalk took up almost none. Apparently the sterile florets were responsible for temperature control in the inflorescence. Subsequent studies confirmed this deduction and showed that the florets do not need the fancy temperature-regulating systems of animals. They contain their own thermostats, their own nutrient supplies and their own means of acquiring oxygen.

In the experiments revealing the existence of the thermostats we removed sterile male florets from the spadix and put individual florets in incubators kept at set temperatures. On their own, the florets could not warm one another and took on the temperature of the air around them. We could therefore assess how much heat they produced at particular temperatures.

The separated florets generated little heat when they were close to freezing, presumably because the enzymes (biological catalysts) needed for heat production—like most enzymes in living creatures—cannot function quickly when they are very cold. But at warmer temperatures, the florets displayed an interesting pattern of heat generation. (That pattern also occurs when the florets are attached to the spadix, except that in the intact spadix the florets gain added warmth from the heat emitted by one another.) As the temperature of the florets rises, so does their rate of heat production, which leads to further warming. This self-reinforcing increase continues until the florets reach 37 degrees C. At higher temperatures, the florets "turn down the furnace," dropping the rate of heat production steeply.

Exactly how far the rate declines depends on the environmental temperature. If the air is cold, say four degrees C, the florets lose heat quickly and their temperature stabilizes at about 38 degrees C (with a high rate of heat generation). On the other hand, when the environment is warm, perhaps 39 degrees C, they lose heat slowly and stabilize at about 46 degrees C (with a low rate of heat production). This pattern is reversible. A floret that has lowered heat production during the warmth of the day can resume generating heat when the temperature drops during the night.

Heat production declines in hot florets probably because the heat itself inhibits the pathways responsible for generating warmth. No one knows whether the high heat acts directly on certain enzymes in the pathways or whether it interferes with enzymatic activity by changing the structure of the membranes that bind the enzymes.

Extraordinary Wattage

Having gained insight into the setting maintained by the florets' thermostat, we began to investigate how the florets obtain the nutrients and oxygen they use during heating and exactly how much heat they produce. It turned out that all the energy devoted to heating in *P. selloum* was present in the florets from the beginning. (This property may not reflect that of other thermoregulating plants, however. The skunk cabbage has to import fuel from the root.) And we were surprised to find that the flowers were "burning" fat, rather than carbohydrate, as had been shown in other aroids. In fact, electron microscopy of the sterile male florets revealed that their tissue contained fat droplets and many mitochondria—in other words, the tissue was remarkably similar to brown fat, a specialized heat-producing tissue found in mammals. Plant and animal cells typically use mitochondria to incorporate into ATP most of the energy derived from nutrients. But in brown fat and apparently in *P. selloum*'s unusual tissue, nutrients and oxygen are used to make heat directly.

P. selloum's impressive ability to produce heat is perhaps best appreciated by comparing the plant's output with that of other plants and animals. A 125-gram spadix produces about nine watts of heat to maintain a temperature of 40 degrees C in a 10 degree C environment—about the same wattage produced by a three-kilogram cat in the same environment. (Because of this correspondence, I often envision *P. selloum* inflorescences as cats growing on stalks.)

On a weight-specific basis, the rate of heat production in *P. selloum* florets approaches the highest rates in flying birds and insects. The florets, which each weigh about eight milligrams, put out 0.16 watt per gram of tissue; the birds and insects emit 0.2 to 0.7 watt per gram. Indeed, when evaluated in this way, aroids in general turn out to be among the greatest heat producers, even if they are compared with animals. The peak performer of the aroids, *A. maculatum*, generates 0.4 watt per gram in its florets, not even an order of magnitude lower than the approximately 2.4 watts per gram output of active flight muscles in bees. This bee muscle rate is the maximum I know for animal tissue.

The high wattage of *P. selloum* raised the question of how it obtains the requisite oxygen, given that it lacks lungs, a circulatory system and the hormones that step up respiration and circulation in animals. We found that the florets gain the oxygen by simple diffusion from the air, which is normally about 21 percent oxygen. Because oxygen levels inside the florets are below the levels in the air, the oxygen moves down the gradient into the plants. Our experiments show that the diffusion of oxygen begins to decrease only after the oxygen concentration around the florets drops below about 17 percent. This level is almost never reached, however, even when the florets are producing heat at their maximum rate.

My recent anatomical studies have defined the pathway through which oxygen penetrates the florets. A floret is about seven millimeters long and 1.2 millimeters thick (roughly the size of an uncooked rice grain). Incredibly, oxygen enters through only about 170 pores, or stomates, and is distributed through a network of spaces that occupies less than 1 percent of the floret's volume. The diffusion path from the floret surface to each cell is somewhat less than 0.75 millimeter, a distance remarkably similar to the length of the air-filled tubes that supply oxygen to insect flight muscles.

Our work with hot plants demonstrates the power of applying ideas and methods developed in one field of science to another field. We found striking similarities between animals and plants, two groups of organisms usually considered to have little in common.

The Voodoo Lily

Bastiaan J.D. Meeuse

Animals differ considerably in their rates of metabolism; a shrew, for example, has a far higher metabolic rate than a sloth. One does not usually think of such differences in plants, but they do exist. The highest rate of metabolism among the higher plants is found in the tissues of certain members of the arum family, which includes such familiar plants as the calla lily, caladium and philodendron. One of the arums—the voodoo lily (*Sauromatum guttatum*)—has such a high rate of metabolism that on the first day of its flowering the temperature of part of it exceeds the temperature of the air by 10 to 15 degrees centigrade (18 to 27 degrees Fahrenheit). The heat volatilizes some of the plant's content of amines and ammonia so that it emits a characteristic unpleasant odor, reminiscent of dung, urine, carrion or decaying blood.

The smell of the voodoo lily and other malodorous members of the arum family serves a useful purpose: it attracts flies and some kinds of scavenger beetles that pollinate the plants. The source of the heat is respiration, which in plants as in animals is the process by which such foodstuffs, as carbohydrates are oxidized in a series of steps catalyzed by enzymes to provide energy for the organism's life processes.

The arum family (formally known as the Araceae) is one of the largest among the monocotyledonous plants; it consists of some 2,000 species in more than 100 genera. Arums live not only on land but also in the water; some are epiphytic; that is, they live on other plants without being parasitic. One of the aquatic arums—water lettuce (*Pistia stratiotes*)—is a rootless plant that spends its life afloat, supported by spongy, air-filled tissue on the bottom surface of its leaves. Another, *Cryptocoryne ciliata*, takes root in the bottom of streams and ponds and in flowering exposes a "flag" on the surface that lures insect pollinators into a foot-long underwater floral chamber.

Both philodendron and the less familiar *Monstera deliciosa* (whose flowers develop into a prized tropical fruit) are epiphytes that climb like vines on the trees of their tropical forest habitat. Of the terrestrial arums perhaps the best-known are the wild forms of the Northern Hemisphere: the skunk cabbages (*Symplocarpus* in eastern Asia and eastern North America and *Lysichitum* in Asia and the American West); jack-in-the-pulpit (*Arisaema*), found in Asia and eastern North America; lords-and-ladies (*Arum maculatum*), its European relative, and another European species known as Italian lords-and-ladies (*Arum italicum*). Some terrestrial species are of considerable economic importance; among them are the sweet flag (*Acorus calamus*) and Indian ipecacuanha (*Cryptocoryne spiralis*), both of which have roots that are used as medicines, and taro (*Colocasia antiquorum*), which is cultivated as a foodstuff throughout the South Pacific.

It is evident from their numbers and diversity that the arums have been highly successful in the evolutionary struggle for survival. One of the reasons for their success is the ingenuity displayed by the ways in which they are pollinated. Cross-pollination, as opposed to self-pollination, confers an evolutionary advantage on a plant species, and even among the most primitive arums, such as the skunk cabbages, self-pollination is unusual. Although the hundreds of tiny flowers that develop on each skunk cabbage's fleshy central spadix are bisexual, their pollen-receiving female organs mature sooner than their pollen-shedding male stamens. Moreover, the stamens tend to mature from the base of the spadix upward, thereby minimizing the likelihood of pollination by gravity. Instead the plants are cross-pollinated by beetles that move from one skunk cabbage to another, attracted by the pungent odor the plants emit when they flower.

The pollination of the Italian lords-and-ladies, one of the more highly evolved arums, is perhaps more typical of the family as a whole. The flowering spadix of *A. italicum* is hot and has an unpleasant smell. The spadix protrudes from the floral chamber like a poker; this upper part is called the appendix. The odor wafting from the appendix on the plant's first day of flowering is a highly effective insect attractant. In 1926 Fritz Knoll, then at the University of Prague, made models of the flowering parts of three European arums and scented the models either with decaying blood or with the appendixes of live arums. He found that his scented models attracted many midges. In nature both the arum's appendix and the inside of the floral chamber are coated with tiny droplets of oil that deny a foothold to the insects that are lured to the plant. The visitors literally fall through the bristles at the top and middle of the chamber. They settle among the female flowers at the bottom of the chamber, where they feed on the sweet fluid exuded by the flowers' stigmas. In the process they also become coated with the fluid.

Overnight the male flowers in the upper part of the floral chamber ripen and shower the insects with golden pollen. Soon the insects, already covered with sticky fluid, are coated with the pollen. When they resume their attempts to escape the next day, they find that the bristles have wilted somewhat and now provide a better foothold. They emerge from the floral chamber and take flight—only to be attracted by the scent of another flowering arum. Then they are captured again and deliver their freight of pollen to the female flowers.

In examining these events from the physiological point of view, one may first ask what purpose is served by the intense metabolism that heats the arum's appendix. Opinion on the matter has been divided. To test whether or not the heat alone is attractive to the pollinating insects. Knoll placed inside his models lightbulbs that made them warmer than the surrounding air. The heated but odorless models failed to draw any insects. Knoll concluded that the sole function of the heating of the appendix was to increase the rate at which the scent compounds were volatilized.

It is evident that the heating of the arum's appendix during flowering constitutes a particularly elegant demonstration of what has been called "the principle of biological parsimony." Heat production and scent production are two aspects of a single respiratory process; both events are simultaneous and inextricably linked. As Knoll suggested, the larger part of the energy serves to volatilize the malodorous compounds, which are made at the expense of very little energy or none at all.

In plants any energy that appears in the form of heat during respiration is normally considered wasted, and in order to avoid such waste the respiratory process depends on the chemical process known as "coupling." Cou-

pling provides that, in several of the enzyme-catalyzed steps involved in respiration, surplus energy is not dissipated as heat but is trapped in the form
of adenosine triphosphate (ATP) or of comparable substances. This coupling between a specific respiratory step and the creation of a high-energy
phosphate is usually obligatory. In the laboratory, however, coupling can be
abolished by the addition of small amounts of such "uncouplers" as azide,
pentachlorophenol or 2,4-dinitrophenol (DNP).

The arums' period of intense respiration is quite brief. The voodoo lily
heats up in a few morning hours and lords-and-ladies in the course of an
afternoon. Both the intensity and the brevity of the process suggest that it is
an example of the uncoupled kind of respiration. One way to test this
assumption is to add an uncoupling agent such as DNP to slices of appendix tissue taken from arum plants at various stages of development. If respiration in the appendix is indeed uncoupled at the time of the plant's
flowering (which we shall call D-day), then the administration of an artificial uncoupler to a bit of naturally uncoupled D-day tissue should not give
rise to a respiration rate significantly different from that of an untreated bit
of D-day tissue. In tissue samples taken at other stages in the plant's development, however, the administration of an uncoupling agent should
enhance respiration in a clear-cut manner.

An experiment conducted by Conrad Hess proved this to be true for
appendix tissue from voodoo lilies. In samples of tissue taken at D-day both
untreated tissue and tissue treated with DNP exhibited about the same respiration rate. When DNP was added to samples taken on the two days following D-day, however, the treated tissue showed a respiration rate at least
double that of the untreated tissue.

Another way to account for the voodoo lily's brief period of runaway
activity is to assume that the plant does not depend on conventional uncoupling to achieve intense respiration. In ordinary respiration the last step of
the process is catalyzed by cytochrome oxidase, an enzyme that can be inactivated by cyanide. In 1938 A. W. H. van Herk of the University of Amster-
dam noted that the administration of cyanide to arum appendix tissue had
no effect on its respiration rate. Van Herk took this to mean that the plants
had a respiratory system that eliminated some of the usual steps in the res-

piratory process. Later studies, however, presented a striking paradox. In spite of the fact that the respiration rate of arum appendix tissue slices is not affected by cyanide, other tests clearly show that the tissue does contain cytochrome oxidase. Thus it would seem that the respiratory steps leading to cytochrome oxidase cannot be absent from the tissue's respiratory process.

Is there any way to resolve the paradox? Britton Chance and his associates at the University of Pennsylvania School of Medicine proposed one way. If the amount of cytochrome oxidase in the appendix tissue is large enough at the start, they suggest, then even when the bulk of the enzyme has been inactivated by cyanide, enough of it will remain unaltered to be able to participate in normal respiration. To account for certain experimental findings on this basis, however, requires the assumption that the appendix tissue contains 60 times the expected amount of cytochrome oxidase. This is not easy to accept.

An alternate resolution of the paradox assumes the existence of a "bypass" system. On this hypothesis most of the respiration in the appendix tissue goes through the usual steps leading to cytochrome oxidase, but these steps can be bypassed under special circumstances, among which is an encounter with cyanide in the laboratory. When this hypothesis is taken out of the laboratory and examined in the voodoo lily's natural environment, a potential evolutionary factor becomes apparent. Any plant equipped with two respiratory pathways would probably not continue to inherit the extra pathway through thousands of generations unless possession of the two pathways had some significant survival value.

Putting aside the question of the voodoo lily's precise mode of respiration, an equally intriguing topic for investigation is the means by which the plant's dramatic D-day activity is triggered. With the aid of electron microscopy E. W. Simon of the University of Manchester detected structural changes in the mitochondria that evidently reflect increased activity as the appendix develops in the period before D-day. In our laboratory we have detected a gradual buildup of flavine compounds and certain biochemical cofactors in appendix tissue in the same period. Neither the structural nor the biochemical changes, however, are sufficient to account for the explosiveness of the D-day flare-up.

In 1939 van Herk made a breakthrough in research on the mechanism of triggering. He cut the male flowers away from the spadix of a voodoo lily more than 18 hours before the plant was due to blossom. The appendix failed to heat up or to produce the characteristic odor. When he delayed the excision of the male flowers until six to 18 hours before blossoming, the appendix became hot and malodorous on schedule. On the basis of these experiments van Herk proposed that the male flower buds secrete a triggering hormone. Later he obtained an extract of the male buds and used it to treat pieces of appendix tissue from voodoo lilies he had rendered inactive by early excision of the male flowers. After a lag of 19 to 22 hours the treated tissues became hot.

Van Herk's experiments clearly showed that the voodoo lily does secrete some substance that triggers the appendix-heating process.

As often happens in studies of this kind, our experiments have both left a number of old questions about the physiology of the arums unanswered and added new questions to the list. Nevertheless, one thing is certain: The plant physiologist and the biochemist have in the appendix tissue of the voodoo lily and comparable members of the arum family an invaluable experimental material.

Carnivorous Plants

Yolande Heslop-Harrison

Green plants gain their energy from the sun, their carbon from the atmosphere and their water and mineral elements from the soil. The atmospheric carbon (in the form of carbon dioxide) and the soil nutrients are replenished from the wastes of microorganisms and animals, and in this way plants and animals are complementary in the general economy of nature. A few plants, however, have evolved the capacity of feeding directly on animals, supplementing their nutrition by capturing and digesting animal prey. By adopting this habit they have gained the ability to survive in nutrient-poor environments, although in some instances at the expense of being able to exist in richer habitats in competition with species that have a more usual life-style.

Those flowering plants that have evolved the carnivorous habit can be divided into two groups according to their methods of catching prey: active trappers and passive trappers. Of the active trappers *Dionaea muscipula*, the Venus's flytrap, is one of the most familiar. In nature this species is found only in certain habitats on the coastal plains of North and South Carolina. Now, however, it is also widely cultivated and can even be seen on sale as a novelty at the checkout counters of supermarkets. Its natural prey are mainly hopping or crawling insects and spiders. Prey touching the leaf agitate tactile hairs; the action triggers a closing mechanism, and the hinged leaf snaps shut.

The commonest of all the active trappers belong to the genus *Utricularia*, which includes some 150 species. In the aquatic or semiaquatic species of this genus the traps take the form of small elastic-walled bladders, hence the common name bladderwort. When the bladder is "set," it is flattened, and the entrance to it is sealed by a flap of cells. The prey are swept into the bladder with the current of water produced when the walls spring apart following the opening of the entrance flap, an action that is triggered by tactile hairs near the entrance.

The passive trappers include the pitcher plants, where the prey is captured and digested in pitcherlike structures formed by modification of the entire leaf, as in the North American genus *Sarracenia*, or by an extension of the leaf tip, as in the genus *Nepenthes* of the eastern Tropics. The prey are enticed to enter the pitcher by colors and scents, much in the way that pollinating insects are attracted to flowers, and they are then drowned and digested in the pitcher fluid. A different strategy is seen in plants with flypaperlike leaves, as in the species of *Pinguicula*, the butterworts, and *Drosera*, the sundews. In these genera glands on the leaf surface secrete adhesive droplets. The prey, usually a flying insect attracted by odor or color or perhaps by the brilliant refractions of the droplets, is trapped by the adhesive when it alights. It becomes more firmly attached to the leaf surface as its efforts to escape bring it into contact with more glands.

The effectiveness of carnivorous plants as predators has been well documented for several species, and the published lists of animal species captured are remarkably long. The prey are usually quite small, but mice have been found in the pitchers of *Nepenthes*, probably the victims of a chance fall, and the remains of small tree frogs have been found in the pitchers of *Sarracenia*. Bladderworts sometimes catch small fishes and tadpoles, but the bladders are never more than a few millimeters wide and are more adapted to the capture of rotifers, copepods and the aquatic larvae of such insects as mosquitoes.

The quantity of prey captured is sometimes quite large. In the pitcher plants the pitchers usually survive for several months, and they may be virtually filled with the decaying remains of their catch. In plants with more ephemeral traps, such as the species of butterworts in which the effective life of a leaf may be only five days, the total catch of a growing season is more difficult to assess. A butterwort such as *Pinguicula grandiflora* grows one new leaf about every five days, so that a total of 400 square centimeters of catching surface may be produced in a single season, even though the diameter of the leaf rosette never exceeds eight centimeters.

What, then, is the value of the carnivorous habit in plants? Charles Darwin, who was one of the pioneers of work on the physiology of carnivorous plants, addressed himself to the question more than a century ago. With his son Francis he showed convincingly that sundews in cultivation that had been fed artificially by applying insects to their leaves were more vigorous, produced more flowers and set more seed than those denied such fare. Richard Harder of the University of Göttingen and others have shown that butterworts, sundews and bladderworts cultivated in controlled environments with carefully regulated access to nutrients perform better when provided with prey, confirming the Darwins' results. They have also found that butterworts make use of pollen carried in the atmosphere, digesting it much as they digest insects.

The nutrients derived from captured prey enter the leaves with surprising rapidity. Some years ago Bruce Knox and I used algal protein labeled with the radioactive isotope carbon 14 in order to follow the movements of the products of digestion in the plant. The leaves of butterworts were fed minute quantities of the labeled protein, and the breakdown products were traced by means of autoradiographs. We found that the amino acid and peptide products of protein digestion moved into the leaf in two or three hours and then passed into the stem and on toward the roots and growing points in less than 12 hours. The main pathway through the leaf was the xylem, the water-conducting tissue of the plant.

John S. Pate and Kingsley Dixon of the University of Western Australia labeled fruit flies by feeding them on yeast containing the isotope nitrogen 15 and have used the flies as a diet for *Drosera*. In the sundews growth arises from corms formed during the preceding season, and much of the nitrogen reserve in the corms is in the form of the amino acid arginine. Pate and Dixon found that at the end of their experiment some 40 percent of the arginine in the corms of the experimental plants contained nitrogen 15, a striking demonstration of how important the supplemental nutrients obtained from the prey must be for the survival of the species in nature.

About a quarter of a million species of flowering plants exist on the earth today, and of them some 400 are known to be carnivorous. They belong to 13 or so genera of six families. Some of the families are quite diverse, with members on every continent. If the main function of the carnivorous habit is to provide scarce nutrients, one might expect the plants to inhabit the kinds of environment where such supplementation would be most beneficial. That is just what is found. The plants are most often encountered in nutrient-poor communities: on healths or in bogs, on impoverished soil in forest openings and occasionally on marl, the crumbly clay soil associated with weathered limestone. Often two or three different genera of carnivorous plants will be found growing together in such localities. For example, in the Pine Barrens of New Jersey several species of sundew, pitcher plant and bladderwort coexist.

At the same time some carnivorous species occupy remarkably narrow ecological niches. Certain bladderworts of South America are found only in the pools of water that accumulate in the natural basins formed by the leaf rosettes of bromeliads, members of the pineapple family. In this environment they are essentially free from competition. Species of the pitcher-plant genus *Heliamphora* provide another example. They occur in nature only in the remote mist zone high in the mountains along the border between Guyana and Brazil, and their environmental requirements are so peculiar that the plants can be maintained in greenhouse cultivation only with difficulty.

Contemplating the range of adaptations found together in the plant carnivores, one can readily appreciate Darwin's absorption with them as examples of evolutionary virtuosity. The trapping mechanisms themselves represent elaborately modified leaves or leaflike organs, and they are usually associated with lures and guides that tempt and direct the prey into or onto the trap. Specialized glands secrete the digestive enzymes, and the same glands or others retrieve the products and pass them back into the plant for distribution through the conducting tissues to the sites of growth. None of the individual features—the traps, lures, odors, directional guides, secreting glands and absorbing glands—is itself peculiar to the carnivores. Many plants have leaf parts capable of rapid movement, for example *Mimosa pudica*, the sensitive plant; others have elaborate insect-trapping mechanisms associated with pollination, and plants of many families have glands capable of secreting water, salt, mucilage, sugars, proteins and other products. It is the assemblage of features that gives carnivorous plants their unique character, the bringing together of so many individual adaptations into a functional combination directed to an end so unusual for a photosynthetic plant that it seems grotesque and even macabre.

Little could be added to the remarkably precise work of these earlier observers until the advent of electron microscopy. The transmission electron microscope has revealed many features of subcellular structure connected with the processes of secretion and absorption in carnivorous plants, and the scanning electron microscope has provided revealing new views of the traps and their associated glands.

The traps of the different types of carnivorous plants have several kinds of surface glands, some concerned with capture and digestion and some with other functions. Certain glands produce nectar as a lure for prey, much as the nectar-secreting glands in flowers that attract insect pollinators do. In *Nepenthes* such glands are around the lip of the pitcher; in *Sarracenia* glands of this type may form an "ant-guiding" trail up the outer surface of the pitcher.

In the butterworts the glands that secrete the viscous globules of the "flypaper" leaf surface are specialized for that function only. In the sundews, however, the stalked glands secrete not only the adhesive but also the enzymes that digest the captured prey. The sundews also possess many minute stalkless glands, which are visible only with the microscope; these glands are scattered over the upper surface of the leaf and on the stalks of the larger glands. The function of the stalkless glands is not known, but it seems possible they are concerned with the movement of the larger stalked glands. The larger glands, which earlier observers saw as "tentacles," move in the direction of the prey when they are stimulated. The movement results from the loss of turgor in groups of cells along the side of the stalk closest to the stimulus. The stalkless glands may be responsible for the withdrawal of fluid that causes the loss of turgor. Glands of a similar type may serve the same function in the butterworts when the leaf rolls up to enfold a captured insect, forming what Darwin called a temporary stomach.

The digestive glands of the carnivorous plants function under different conditions in the various genera according to the nature of the trapping mechanism. In *Nepenthes* the glands in the lower third of the pitcher become totally immersed in their own secretion fluids as the trap matures and before any prey is captured. In the larger species the accumulation of fluid can be as much as a liter. In the four other genera of pitcher plants smaller quantities of fluid collect, in some cases scarcely enough to immerse the glands, but here again it seems that the presence of prey is not necessary to stimulate secretion.

In contrast, the digestive glands on the leaf of the Venus's flytrap remain dry until the prey is captured. If the trap is sprung with a pencil or a glass rod, the digestive glands remain dry and the leaf soon reopens. When an insect is trapped, however, the glands become active and a secretion pool builds up between the closed lobes of the leaf. Evidently the onset of secretion depends on chemical stimulation rather than mechanical. Although the glands in the bladders of the bladderworts are permanently immersed in water, it seems that they too do not secrete enzymes until they are stimulated by prey.

The sundews are different. Here the viscid secretion droplets accumulate on each gland head as the leaf matures; the load is held until the gland is touched by prey, and then still more secretion is released. The butterworts and the dewy pine *Drosophyllum* in some respects combine the features of the Venus's flytrap and the sundews. In these genera there are two classes of glands on the leaf surface: stalked glands that carry secretion droplets at maturity and are mainly concerned with catching prey, and stalkless ones that remain dry until they are stimulated, when they pour out a less viscid secretion containing the digestive enzymes.

In Darwin's experiments on the sundews and butterworts he found that secretion could be stimulated by many sources of combined nitrogen but not, for example, by sugar or sodium carbonate. Insects excrete many nitrogenous products, but the main stimulant may be uric acid, which is abundantly present in all insect excreta.

The digestive glands of the carnivorous genera vary considerably in their morphology. In the pitcher-plant genus *Nepenthes* the glands are some 60 micrometers in diameter and are partly sunk below the inner epidermis of the pitcher, where they are protected by an overlying flap of tissue. In the sundews the heads of the digestive glands are borne on multicellular stalks. The glands of the butterworts, both the stalked glands specialized for insect capture and the sessile ones concerned with digestion, are smaller and consist of many fewer cells than the glands of the other genera.

Notwithstanding such structural variations, a common architectural theme can be traced in all the digestive glands. Indeed, the theme is one that recurs in many other classes of plant-surface glands. In all cases it is the secretory cells that form an outer cap or layer one cell or a few cells thick, which lies directly over a specialized single cell or a pavement composed of several such cells side by side. This second layer is either in direct contact with the conducting vessels of vascular tissue or is separated from such tissue by two or three large "reservoir" cells.

The secretory outer cells of the gland are epidermal cells specialized for their function of enzyme synthesis, and they show many features reminiscent of those found in animal cells with similar functions. The network of cytoplasmic membranes known as the endoplasmic reticulum is well developed, and sometimes the elements are stratified, as they are in the cells of the animal pancreas.

The secretory cells of carnivorous plants are comparable to those of animals in still other ways. The vacuoles of the secretory cells, formed as inflated bays of the endoplasmic reticulum, are sites of enzyme storage: they are therefore comparable to the lysosomes of animal cells. Furthermore, in some instances the nuclei of the secretory cells of the gland head contain more DNA than most body cells. This is a feature of such animal glands as the salivary gland of the fruit fly.

The cells of animal glands have an outer membrane but do not have a cell wall of the kind found in plants, and many of the adaptations of the gland cells of the carnivorous plants are unique in that they have to do with the structure and function of the cell wall. The outer walls of the secretory cells are coated with a water-resistant, cutinized layer, but this layer is perforated by distinct pores or less well-defined discontinuities through which the secretions can reach the outer surface of the cell.

The digestive glands secrete enzymes by other methods. In some instances the enzymes seem to diffuse directly through the plasmalemma, the outer membrane of the cytoplasm. In others, as in the sundews, the transfer involves a local disruption of the plasmalemma during the period of rapid secretion that follows the capture of the prey.

The cells of the layer underlying the secretory cells show some of the characteristics of those of the endodermal layer of the root, a sheath of cells that separates the root cortex from the inner conducting tissues. The side walls of the cells are heavily cutinized, and in them the plasmalemma is fused with the cell wall. Water cannot pass through the side walls, and so it is constrained to move through the cytoplasm.

Because the glands of the different genera of carnivorous plants function under widely different circumstances it is only to be expected that the processes of secretion and resorption should vary accordingly. My own observations of such "flypaper" trappers as butterworts and sundews suggest that these plants have secretory and resorptive mechanisms that are quite different from the ones likely to operate in pitcher plants. Among the butterworts some enzymes, notably amylases, are secreted by the stalked glands whose sticky exudate captures the insect prey, but it is the stalkless glands at the surface that furnish the main outflow of digestive fluid. Before stimulation the stalkless glands hold in reserve a supply of proteases, nucleases, phosphatases, esterases and other digestive enzymes, stored both in the spongy cell walls and in the vacuoles of the secretory cells. Stimulation induces the stored enzymes out onto the surface of the leaf.

One can follow the activity of the enzymes that build up in the pool of secretion on the leaf surface. The pool extends and deepens to engulf the prey, and then, after the digestion is completed, the fluid is resorbed. Generally speaking, the size of the pool is related to the size of the prey. A small captive insect induces only a modest flow of digestive fluid, and in the butterworts after such a catch the pool may reach its maximum size in an hour or so. A large insect may stimulate so much secretion that surplus fluid will drip off the edges of the leaf. Under such circumstances the secretion may go on for several hours, and its volume may exceed the entire volume of liquid originally held in the leaf, showing that the flow is supplemented by the passage of water from elsewhere in the plant through the vascular system. Overstimulated leaves do not complete the digestive cycle. Resorption does not take place, and the leaf begins to rot, a victim, so to speak, of plant indigestion.

The butterworts, and probably other genera of carnivorous plants with the same pattern of digestive-gland function, have thus evolved a definite digestive cycle. The secretion and resorption phases, respectively associated with massive movements of fluid first outward and then inward through each gland, are geared to far-reaching changes in the gland cells. There appears to be no such cycle in the pitcher plants. In this group the early period of secretion is followed by a prolonged interval when the pitcher holds a more or less constant amount of fluid. Trapped insects accumulate in the fluid, and the digestion products are withdrawn continuously from the pitcher into the main body of the plant.

The uptake mechanism proposed for the *Nepenthes* pitcher is distinctly similar to the mechanism assumed to be responsible for the normal uptake of soil minerals by plant roots. It is as though in each pitcher the plant were creating its own enriched soil solution and abstracting from it the minerals it needs. The analogy seems even apter when one considers that after the pitcher has been open for some time its fluid becomes infected with a commensal flora, mostly bacteria, that quickly assumes most of the burden of digesting the captured prey.

At this stage the pitcher fluid has become distinctly alkaline—and distinctly malodorous. The plant enzymes may now play little part in the digestive process; the digestive glands act mainly as organs of absorption, selectively taking up and concentrating useful products. To carry the analogy even further, it is possible that pitcher plants rooted in the ground may benefit at the beginning of each season's growth from a temporary local enrichment of the soil by nutrients released by the decay of the preceding season's dead pitchers and their partially digested contents. Here the useful products of predation would be taken up in the usual plant fashion: through the roots rather than the leaves.

It seems clear that the supplemental nutrients available to carnivorous plants offer them special advantages, particularly in environments where certain kinds of nutrients are scarce. It has commonly been supposed the principal benefit of the capture and digestion of animal prey by a plant is a supplemental supply of nitrogen. Current research indicates, however, that supplementary phosphorus is equally important and perhaps in some circumstances even more important. The presence in the digestive-gland secretions of nuclease and phosphatase enzymes may well be related to this requirement. In habitats where plant growth is limited by deficiencies of major nutrient elements such as phosphorus—or of one or more of the other elements required only in trace amounts—the advantages to be gained by acquiring contributions from animal prey would be substantial.

So much for the advantages of the carnivorous habit. Are there counterbalancing costs? Most plants live in competitive circumstances; is the carnivorous plant's energetic investment in the synthesis of digestive enzymes and other secretion products, not to mention the investment in the plant's elaborate structural adaptations, cost-effective? The intriguing conclusion of this line of thought is simply that any energy balance sheet is scarcely relevant. In all but a few instances the carnivorous plants are found in places where an abundance of sunlight, adequate carbon sources and unlimited access to water during the growing period place no limit on photosynthesis, the primary energetic resource of the plant. Thus the energetic cost of capturing an atom of nitrogen or of phosphorus, or of whatever else may be the principal growth-limiting element, is not significant. If the capture of such vital nutrients enables the plant to survive in places where no noncarnivorous competitor can intrude, then it is proved that, whatever the energetic cost may be, the investment is justified.

Suggested Reading

Meeuse, B. J. D., and I. Raskin. "Sexual Reproduction in the Arum Lily Family, with Emphasis on Thermogenicity," *Sexual Plant Reproduction* 1, no. 1 (March 1988): 3–15.

Seymour, R. S., G. A. Bartholomew, and M. C. Barnhart. "Respiration and Heat Production by the Inflorescence of *Philodendron selloum,*" *Planta* 157, no. 4 (April 1983): 336–343.

Seymour, R. S., M. C. Barnhart, and G. A. Bartholomew. "Respiratory Gas Exchange During Thermogensis in *Philodendron selloum,*" *Planta* 161, no. 3 (June 1984): 229–232.

Seymour, R. S. "Analysis of Heat Production in a Thermogenic Arum Lily, *Philodendron selloum* by Three Calorimetric Methods," *Thermochimica Acta* 193 (December 14, 1991): 91–97.

Seymour, R. S., and P. Schultze-Motel. "Thermoregulating Lotus Flowers," *Nature,* 383 (September 26, 1996): 305.

Chapter 7

Small but Deadly

The Toxins of Cyanobacteria

Wayne W. Carmichael

On May 2, 1878, George Francis of Adelaide, Australia, published the first scholarly description of the potentially lethal effects produced by cyanobacteria—the microorganisms sometimes called blue-green algae or, more colloquially, pond scum. In a letter to *Nature* he noted that an alga he thought to be *Nodularia spumigena* had so proliferated in the estuary of the Murray River that it had formed a "thick scum like green oil paint, some two to six inches thick, and as thick and pasty as porridge." This growth had rendered the water "unwholesome" for cattle and other animals that drink at the surface, bringing on a rapid and sometimes terrible death.

Since 1878, investigators have confirmed that *Nodularia* and many other genera of cyanobacteria include poisonous strains. Indeed, such microbes are known to account for spectacular die-off of wild and domestic animals. In the midwestern U.S., for instance, migrating ducks and geese have perished by the thousands after consuming water contaminated by toxic cyanobacteria. In recent years, workers have identified the chemical structure of many cyanobacterial toxins and have also begun to decipher the steps by which the poisons can lead to suffering and death.

Such research is exciting interest today, in part because of worry over public health. No confirmed human death has yet been attributed to the poisons. But runoff from detergents and fertilizers is altering the chemistry of many municipal water supplies and swimming areas, increasing the concentration of nitrogen and phosphorus. These nutrients promote reproduction by dangerous cyanobacteria and thus foster formation of the dense growths, known as waterblooms, described by Francis. As cyanobacterial waterblooms become more common in reservoirs, rivers, lakes and ponds, the likelihood grows that people will be exposed to increased doses of toxins. (Water-treatment processes only partially filter out cyanobacteria and dilute their toxins.) The risk of animal die-offs grows as well.

The possibility of increased exposure has become particularly disturbing because some evidence suggests that certain cyanobacterial toxins might contribute to the development of cancer. Knowledge of the chemical structure and activity of the toxins should help scientists to devise more sensitive ways to measure the compounds in water and to develop antidotes to lethal doses. Improved understanding of how these chemicals function should also facilitate efforts to determine the long-term effects of exposure to nonlethal doses.

Research into the structure and activity of the toxins is sparking interest on other grounds as well. They and their derivatives are being considered as potential medicines for Alzheimer's disease and other disorders. The substances already serve as invaluable tools for exploration of questions in cell biology.

As worrisome and wonderful as the toxins are, other aspects of cyanobacteria are perhaps more familiar to many people. For example, textbooks often feature these bacteria as nitrogen fixers. The filamentous species (which consist of individual cells joined end to end, like beads on a string) convert atmospheric nitrogen into forms that plants and animals can use in their own life processes. In this way, they fertilize agricultural land throughout the world, most notably rice paddies, where they are often added to the soil.

Cyanobacteria are known, too, for the critical insights they have provided into the origins of life and into the origins of organelles in the cells of higher organisms. The fossil record shows that cyanobacteria already existed 3.3 to 3.5 billion years ago. Because they were the first organisms able to carry out oxygenic photosynthesis, and thus to convert carbon dioxide into oxygen, they undoubtedly played a major part in the oxygenation of the air. Over time, their exertions probably helped to create the conditions needed for the emergence of aerobic organisms. At some point, theorists suggest, certain of the photosynthesizers were taken up permanently by other microbes. Eventually these cyanobacteria lost the ability to function independently and became chloroplasts: the bodies responsible for photosynthesis in plants.

Cyanobacteria were typically referred to as blue-green algae because of the turquoise coloring of most blooms and the similarity between the microbes and true algae (both carry out photosynthesis). But Roger Y. Stanier, then at the University of California at Berkeley, was beginning to reveal the "algae" part of the name to be a misnomer.

After the electron microscope was introduced in 1950, work by Stanier and others established that two radically different types of cells exist in the contemporary world. Prokaryotic varieties—those bearing the characteristics of bacteria—have no membrane enveloping their nuclear material and usually lack membrane-bound bodies in their interior. All other cells, including those of algae and more complex plants, are eukaryotic: they contain a definite nuclear membrane and have mitochondria as well as other organelles. Stanier's subsequent examinations of cyanobacteria prompted him to note in 1971 that "these organisms are not algae; their taxonomic association with eukaryotic groups is an anachronism. . . . Blue-green algae can now be recognized as a major group of bacteria."

Reports implicating the microorganisms in the poisoning of wild and domestic animals had accumulated from many parts of the world. The animals died after drinking from ponds or other waters partly covered by slimy carpets of what seemed to be algae, often in the dog days of late summer and early fall, when the temperature is high and the air is relatively still. Yet no firm link between specific genera of cyanobacteria and animal deaths had yet been established.

Theodore A. Olson, a microbiologist at the University of Minnesota, made that connection in the course of studies he carried out between 1948 and 1950. Olson collected samples of waterblooms in his state and determined that they contained copious amounts of species from the cyanobacterial genera *Microcystis* and *Anabaena* (common groups of planktonic cyanobacteria). By feeding cyanobacteria from those blooms to laboratory animals, he was able to demonstrate that certain water-dwelling forms can indeed be poisonous to animals.

This finding in turn raised new questions. Why, for example, were animals poisoned most often during the dog days of summer and fall? The answer seems to be that cyanobacteria grow remarkably well and form waterblooms when four conditions converge: the wind is quiet or mild, and the water is a balmy temperature (15 to 30 degrees Celsius), is neutral to alkaline (having a pH of 6 to 9) and harbors an abundance of the nutrients nitrogen and phosphorus. Under such circumstances, cyanobacterial populations grow more successfully than do those of true algae. (True algae can also form waterblooms, but blooms in nutrient-rich water usually consist of toxic cyanobacteria.)

Because the cells release toxins only when they die or become old and leaky, animals usually have to ingest whole cells to be affected. They can, however, take in a fatal dose of toxins from cell-free water if someone has treated the water with a substance, such as copper sulfate, designed to break up waterblooms. The amount of cyanobacteria-tainted water needed to kill an animal depends on such factors as the type and amount of poison produced by the cells, the concentration of the cells, as well as the species, size, sex and age of the animal. Typically, though, the required volume ranges from a few millimeters (ounces) to several liters (a few gallons). Apparently, thirsty animals are often undeterred by the foul smell and taste of contaminated water.

The early demonstration that cyanobacterial toxins were responsible for animal kills in Minnesota also raised questions that Gorham took up in the 1950s—namely, what is the chemical nature and modus operandi of the toxins?

In 1972, Carol S. Huber and Oliver E. Edwards determined the chemical structure of a cyanobacterial toxin for the first time. Named anatoxin-a, it turned out to be an alkaloid—one of thousands of nitrogen-rich compounds that have potent biological, usually neurological, effects. Species from seven of 12 cyanobacterial genera involved in animal deaths have been cultured. Interestingly, none of the 12 genera grow attached to rocks or vegetation; all are planktonic, floating in water as single cells or filaments. Most produce more than one type of toxin.

The toxins that have been studied intensively to date belong to one of two groups, defined by the symptoms they have produced in animals. Some, such as anatoxin-a, are neurotoxins. They interfere with the functioning of the nervous system and often cause death within minutes, by leading to paralysis of the respiratory muscles.

Other cyanobacterial poisons, such as those produced by Francis's *N. spumigena*, are hepatotoxins. They damage the liver and kill animals by causing blood to pool in the liver. This pooling can lead to fatal circulatory shock within a few hours, or, by interfering with normal liver function, it can lead over several days to death by liver failure.

Four neurotoxins have been studied in detail. Of these, anatoxin-a and anatoxin-a(s) seem unique to cyanobacteria. The other two—saxitoxin and neosaxitoxin—arise in certain marine algae as well. I had the good fortune of being able to explore the activity of anatoxin-a soon after its structure was deciphered. This compound, made by various strains of the freshwater genera *Anabaena* and *Oscillatoria*, mimics the neurotransmitter acetylcholine.

When acetylcholine is released by neurons (nerve cells) that impinge on muscle cells, it binds to receptor molecules containing both a neurotransmitter binding site and an ion channel that spans the cell membrane. As acetylcholine attaches to the receptors, the channel opens, triggering the ionic movement that induces muscle cells to contract. Soon after, the channel closes, and the receptors ready themselves to respond to new signals. Meanwhile an enzyme called acetylcholinesterase degrades the acetylcholine, thereby preventing overstimulation of the muscle cells.

Anatoxin-a is deadly because it cannot be degraded by acetylcholinesterase or by any other enzyme in eukaryotic cells. Consequently, it remains available to overstimulate muscle. It can induce muscle twitching and cramping, followed by fatigue and paralysis. If respiratory muscles are affected, the animal may suffer convulsions (from lack of oxygen to the brain) and die of suffocation. Unfortunately, no one has succeeded in producing an antidote to anatoxin-a. Hence, the only practical way for farmers or other concerned individuals to prevent deaths is to recognize that a toxic waterbloom may be developing and to find an alternative water supply for the animals until the bloom is eliminated.

For animals, anatoxin-a is an anathema, but for scientists it is a blessing. As a mimic of acetylcholine, anatoxin-a makes a fine research tool. For example, because it resists breakdown by acetylcholinesterase, the toxin and its derivatives can be used in place of acetylcholine in experiments examining how acetylcholine binds to and influences the activity of acetylcholine receptors (especially the so-called nicotinic acetylcholine receptors in the peripheral and central nervous system).

Edson X. Albuquerque and his colleagues at the University of Maryland School of Medicine are looking at anatoxin-a in other ways as well. The researchers are in the early stages of exploring the intriguing possibility that a modified version might one day be administered to slow the mental degeneration of Alzheimer's disease. In many patients, such deterioration results in part from destruction of neurons that produce acetylcholine. Acetylcholine itself cannot be administered to replace the lost neurotransmitter because it disappears too quickly. But a version of anatoxin-a that has been modified to reduce its toxicity might work in its place. Derivatives of anatoxin-a could also conceivably prove useful for other disorders in which acetylcholine is deficient or is prevented from acting effectively, such as myasthenia gravis (a degenerative disorder that causes muscle weakness).

The other neurotoxin unique to cyanobacteria, anatoxin-a(s), is made by strains of *Anabaena*. It produces many of the same symptoms as anatoxin-a—which is how it came to have such a similar name. The letter "s" was appended because anatoxin-a(s) seemed to be a variant of anatoxin-a that caused vertebrates to salivate excessively. However, my students and I at Wright University, together with Shigeki Matsunaga and Richard E. Moore of the University of Hawaii, have shown that anatoxin-a(s) differs chemically from anatoxin-a and elicits symptoms by other means.

Anatoxin-a(s) is a naturally occurring organic phosphate that functions much like synthetic organophosphate insecticides, such as parathion and malathion. To my knowledge, it is the only natural organophosphate yet discovered. Even though its structure differs from that of the synthetic compounds, its killing power, like theirs, stems from its ability to inhibit acetylcholinesterase. By impeding acetylcholinesterase from degrading acetylcholine, it ensures that acetylcholine remains continuously available to stimulate—and overstimulate—muscle cells.

As a structurally novel organophosphate, anatoxin-a(s) could in theory form the basis for new pesticides. Synthetic organophosphates are widely used because they are more toxic to insects than to humans. They are, however, under some fire. Soluble in lipids (fats), they tend to accumulate in cell membranes and other lipid-rich parts of humans and other vertebrates. Anatoxin-a(s), in contrast, is more soluble in water and, hence, more biodegradable. So it could be safer. On the other hand, it might also be less able to cross the lipid-rich cuticles, or exoskeletons, of insects. By tinkering with the structure of anatoxin-a(s), investigators might be able to design a compound that would minimize accumulation in tissues of vertebrates but continue to kill agricultural pests.

As is true of anatoxin-a and anatoxin-a(s), the neurotoxins saxitoxin and neosaxitoxin disrupt communication between neurons and muscle cells. But they do so by preventing acetylcholine from being released by neurons. In order to secrete acetylcholine or other neurotransmitters, neurons must first generate an electrical impulse. Then the impulse must propagate along the length of a projection called an axon—an activity that depends on the flow of sodium and potassium ions across channels in the axonal membrane. When the impulse reaches an axon terminal, the terminal releases stores of acetylcholine. Saxitoxin and neosaxitoxin block the inward flow of sodium ions across the membrane channels; in so doing, they snuff out any impulses and forestall the secretion of acetylcholine.

Cyanobacterial neurotoxins, then, are both deadly and potentially valuable, but they are not as ubiquitous as the other major class of cyanobacterial poisons: the hepatotoxins. Whereas neurotoxins have been blamed for kills mainly in North America (with some in Great Britain, Australia and Scandinavia), hepatotoxins have been implicated in incidents occurring in virtually every corner of the earth. For this reason, great excitement ensued in the early 1980s, when a group headed by Dawie P. Botes, then at the Council for Scientific and Industrial Research in Pretoria, determined the chemical structure of a liver toxin. Such toxins were long known to be peptides (small chains of amino acids), but the technological advances needed for determining the precise structures of the toxins did not occur until the 1970s.

Soon after Botes established the chemical identity of the first few hepatotoxins, my laboratory and others confirmed his results and began identifying the chemical makeup of other hepatotoxins. Extensive structural analyses, mainly in the laboratory of Kenneth L. Rinehart of the University of Illinois, have now established that the liver toxins form a family of at least 53 related cyclic, or ringed, peptides. Those consisting of seven amino acids are called microcystins; those consisting of five amino acids are called nodularins. The names reflect the fact that the toxins were originally isolated from members of the genera *Microcystis* and *Nodularia*.

Research into the hepatotoxins—much of which is carried out at other laboratories with toxins supplied by my group—is directed primarily at understanding how the compounds affect the body. Investigators know that the peptides cause hepatocytes, the functional cells of the liver, to shrink. In consequence, the cells, which are normally packed tightly together, separate. When the cells separate, other cells forming the so-called sinusoidal capillaries of the liver also separate. Then the blood carried by the vessels seeps into liver tissue and accumulates there, leading to local tissue damage and, often, to shock.

Scientists have wondered why the toxins act most powerfully on the

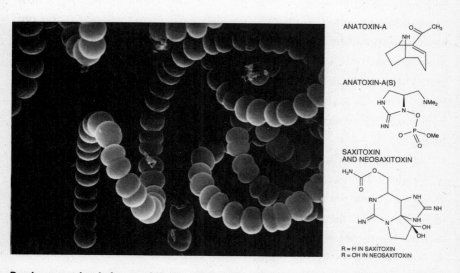

ANATOXIN-A

ANATOXIN-A(S)

SAXITOXIN
AND NEOSAXITOXIN

R = H IN SAXITOXIN
R = OH IN NEOSAXITOXIN

Beads on a string (micrograph) are actually cells of cyanobacterium *Anabaena flos-aquae*, magnified some 2,500 times. *A. flos-aquae* is a major producer of neurotoxins, poisons that interfere with the functioning of the nervous system. The strain shown here was responsible for the death of hogs in Griggsville, Illinois. The chemical structures at the right represent toxins made by strains of *Anabaena*: all except anatoxin-a(s) also occur in other cyanobacteria.

liver. The answer probably is that they are moved into hepatocytes by the transport system, found only in hepatocytes, that carries bile salts into the cells.

The neurotoxins and hepatotoxins are certainly the most dangerous cyanobacterial compounds, but they are by no means the only bioactive chemicals made by these bacteria. For example, the microbes produce an array of cytotoxins: substances that can harm cells but do not kill multicellular organisms. Studies indicate that several cytotoxins show promise as killers of algae and bacteria. Some may even serve as agents for attacking tumor cells and the human immunodeficiency virus, the cause of AIDS.

To what purpose are all these chemicals made? They may enhance cyanobacterial defense against attack by other organisms in the everyday environment. But why would cyanobacteria produce substances capable of killing livestock and other large animals? After all, livestock have never been the primary predators of those microbes.

It turns out that cyanobacterial neurotoxins and hepatotoxins can be extremely harmful not only to birds, cows, horses and the like but also to the minute animals (zooplankton) living in lakes and ponds. The toxins may be directly lethal (especially the neurotoxins), or they may reduce to the number and size of offspring produced by the creatures that feed on cyanobacteria. In other words, just as vascular plants make tannins, phenols, sterols and alkaloids to defend against predation, it is likely that cyanobacteria synthesize poisons to ward off attack by fellow planktonic species.

In support of this suggestion, we have found that zooplankton species generally do not eat cyanobacteria capable of producing toxins unless there is no other food around; then, they often attempt to modulate the amount they take in to ensure that they avoid a lethal dose. Those who walk this aquatic tightrope successfully pay a price, of course, in fewer offspring, but at least they survive to reproduce.

It is possible, though, that the protective effect is incidental. The toxins may once have had some critical function that they have since lost. This likelihood is suggested by the fact that microcystins and nodularins act on the protein phosphatases that regulate the proliferation of eukaryotic cells.

The hepatotoxins do not now seem to participate in cell function and cell division in cyanobacteria, but they may have played such a role early in the evolution of these organisms (and of other microbes as well).

Regardless of their intended purpose, the toxicity of many chemicals produced by cyanobacteria is undeniable. For this reason, I am becoming increasingly worried by a modern fad: the eating of cyanobacteria from the genus *Spirulina* as a health food.

Spirulina itself is not harmful. The danger arises because there are no guidelines requiring those marketing *Spirulina* to monitor their products for contamination by potentially toxic cyanobacteria or by cyanobacterial toxins. Moreover, the general public is ill equipped to distinguish *Spirulina* and other benign cyanobacterial products from poisonous forms of cyanobacteria.

My worry has recently intensified because the popularity of *Spirulina* has led to the production and marketing of such cyanobacteria as *Anabaena* and *Aphanizomenon*—genera that contain highly toxic strains. Some promotional material for cyanobacteria-containing products even claims that the items being sold can moderate some disease symptoms, including those of debilitating neuromuscular disorders. Yet this literature does not provide a listing of all microbial species in the marketed products, nor does it indicate that anyone is monitoring the products to ensure they are pure and nontoxic. Because cyanobacteria are often collected simply from the surface of an open body of water and because neither sellers nor buyers can distinguish toxic from nontoxic strains without applying sophisticated biochemical tests, the safety of these items is questionable.

All told, the cyanobacteria constitute a small taxonomic group, containing perhaps 500 to 1,500 species. But their power to harm and to help animals and humankind is great. Investigated and exploited responsibly, they can provide valuable tools for basic research in the life sciences and may one day participate in the treatment of disease.

The Lurking Perils of *Pfiesteria*

JoAnn M. Burkholder

On a hot, humid October afternoon in 1995, I stood in a gently rocking boat, watching hundreds of thousands of bloody, dying fish break the mirrorlike surface of North Carolina's Neuse Estuary, where the Neuse River mixes with salty water from the Atlantic Ocean. Rising up out of the river, writhing, the fish gasped for air, then became still, floating on their sides. They were mostly Atlantic menhaden, small fish that serve as food for many larger species valued by commercial fishermen. An occasional flounder, croaker or eel also bobbed on the surface. Seagulls lined the shores of the nearly eight square miles of kill zone; a feast was in the making.

With my team from North Carolina State University (N.C.S.U.), I was collecting water samples from the area to try to determine the cause of the deaths. The bloody sores on the fish and their erratic behavior signaled a possible toxic outbreak of *Pfiesteria piscicida*, a single-celled microorganism that we had first seen in 1989 and had later linked to fish kills in several major estuaries. By the time this kill was over, 15 million silvery carcasses would carpet the water.

We quickly completed our sampling and pulled anchor, knowing it would be unwise to linger if *P. piscicida* was the culprit (as our test results later indicated was the case). People who have had contact with this creature in its toxic state have suffered from a range of symptoms, among them nausea, respiratory problems and memory loss so severe that it sometimes has been mistaken for Alzheimer's disease.

The scene on the river was all too familiar. In 1991 a billion fish died in the same way in this estuary. Since then, *P. piscicida*, occasionally with a closely related but unnamed toxic species, has been implicated almost yearly in massive fish kills in the estuaries of North Carolina (where it typically wipes out hundreds of thousands to millions of fish in a year) and in several smaller kills involving thousands of fish in Maryland waters of Chesapeake Bay.

These two species are the first members of the "toxic *Pfiesteria* complex," referred to hereafter as simply *Pfiesteria*. They (or still other toxic species that look the same but have not yet been identified definitively) have now been found as well in coastal waterways extending from Delaware to the Gulf Coast of Alabama, although they have not been linked to fish deaths outside North Carolina and Maryland.

My colleagues and I have learned a great deal about *Pfiesteria*'s life cycle and about the reasons for its proliferation and toxic outbreaks. We have also found it to be an astonishing creature, displaying properties never before seen in dinoflagellates—the larger group of microorganisms to which it belongs. Dinoflagellates, encompassing thousands of species, gain their name from the whiplike appendages (flagella) that they use for swimming in certain of their life stages.

Other unexpected findings have prompted us to look beyond the floating dead fish to *Pfiesteria*'s additional untoward actions. Disturbingly, we have seen that aside from killing many fish at once, *Pfiesteria* can impair the health of finfish and shellfish in more subtle ways, such as by undermining their ability to reproduce and resist disease. These less obvious effects could potentially deplete fish populations more permanently than acute kills do.

Pfiesteria is not alone in its quiet treachery. Work by many investigators has also turned up insidious activities of other "harmful algae." As the term implies, this eclectic category encompasses certain true algae—primitive plants that make chlorophyll and carry out photosynthesis to make their own food. But it also includes various (usually unicellular) creatures, such as *Pfiesteria*, that look like algae but are not plants at all. The members of this ragbag group can hurt fish when they bloom, or proliferate—doing damage by producing dangerous levels of toxins or by other means, such as by growing so extensively that they rob the water of oxygen and cause fish to suffocate.

Various harmful algae are infamous for causing huge fish kills and for acutely poisoning animals or people who ingest toxin-laden seafood or water. Indeed, some of *Pfiesteria*'s dinoflagellate cousins account for the extraordinary red tides that have discolored and poisoned coastal waters worldwide for thousands of years. Yet the less obvious effects of harmful algae also need to be clarified and addressed if other serious illnesses and death in fish—and possibly in humans and other organisms—are to be avoided.

Pfiesteria was first linked to the death of fish in 1988, when tank after tank of fish in brackish water at N.C.S.U.'s College of Veterinary Medicine began dying mysteriously. The veterinarians noticed a swimming microorganism in the water and deduced through microscopy that it was a dinoflagellate. They subsequently noted that it became abundant in the aquarium cultures just before the fish died and seemed to disappear soon after the fish perished. But it reappeared if live fish were added to the tanks.

Because fish from around the world are studied at this laboratory, no one knew where the organism had come from or if it was a species already known to science. In 1989 the veterinarians asked my research group in the N.C.S.U. department of botany to help identify the microbe and determine whether it was responsible for the fish deaths.

The Nature of the Adversary

We soon realized that the creature was unique among both toxic and nontoxic dinoflagellates in adopting some forms, or stages, that do not resemble those of other dinoflagellates at all; in those stages it looks like a group of microorganisms called chrysophytes. It also stood alone among the small subset of dinoflagellates that are toxic. Those species (totaling about 60) produce some of the most potent poisons ever discovered in nature, although they make them for no obvious purpose. But the newfound organism not only appeared to poison fish—it ate them as well!

My research team learned that the extraordinary microbe we eventually named *Pfiesteria piscicida* is nontoxic when fish are absent. When it senses fish excrement and secretions in the water, however, it both emits toxins and swims directly toward the fish materials. The toxins strip away the skin of the fish, damage their nervous system and vital organs and make them too lethargic to flee. Then the fish commonly sustain attacks by other destructive microbes, and bleeding sores develop where the skin has been destroyed. With the fish unable to escape, the dinoflagellate cells feed on sloughed skin, blood and other substances leaking from the sores. Later the lethal cells change from flagellated, swimming forms to more amorphous amoebae that dine on the victims' remains, sometimes becoming so engorged that they can no longer move.

Toxic *P. piscicida* can be a very effective killer. In laboratory tests, toxin-contaminated water or cultures of the cells have killed many finfish and shellfish species. My research associate, Howard B. Glasgow, Jr., has found that young animals, as well as adults of more sensitive species, can die minutes after exposure, and most victims die within hours.

We also discovered another trait that had never been found in other toxic dinoflagellates. Remarkably, *P. piscicida* can transform into at least 24 distinct stages over the course of its life cycle. It alters its shape and size according to available food sources, which include prey ranging from bacteria all the way up the food chain to mammalian tissue. Some of these stage changes can involve a more than 125-fold increase in size and can take place in less than 10 minutes.

We studied *Pfiesteria* for two years in aquarium tanks without knowing where it might have come from. But the information we gathered indoors prepared us for that search. We began by looking in our own "backyard." Every year since at least the mid-1980s, massive fish kills had plagued North Carolina's Albemarle-Pamlico Estuarine System, which contains the Neuse River. With help from state biologists, we obtained water samples in 1991 during a kill of about one million Atlantic menhaden in the Pamlico Estuary.

The Adversary in Nature

When we examined the samples with a scanning electron microscope, we saw small dinoflagellates that looked identical to those we had found in the contaminated vet-school aquariums. Moreover, just as had happened in our tanks, the cells seemed to disappear after the kill ended—they were absent from water samples collected among the floating remains of fish one day after the fish died. This work not only tracked the vet-school contaminant to its probable origin but also implicated *Pfiesteria* as an important cause of fish death in nature.

What triggers toxic outbreaks of *Pfiesteria*? Laboratory and field experiments by many researchers indicate that, among other factors, an overabundance of nutrients such as nitrogen and phosphorus in the water help to set the stage for these events. The shallow, slow-moving waters of many North Carolina estuaries are easily polluted by materials from the surrounding land. These include nutrient-rich human sewage, fertilizers, certain industrial by-products (including some rich in phosphates) and animal wastes (from many swine and poultry operations in the watershed). When the waters become overnutrified, algae proliferate, much as houseplants grow much better when their soil contains added fertilizer. The abundant algae provide a rich food source for *Pfiesteria*, which then reproduces rapidly, creating legions ready to attack schools of fish should they swim into *Pfiesteria*-infested waters.

The estuaries of North Carolina turn out to be a very troubling place for *Pfiesteria* to wreak havoc. The Albemarle-Pamlico is the second-largest U.S. estuarine system outside Alaska, and it provides half the area used by fish from Maine to Florida as nursery grounds. Many young fish come to these waters to grow and develop before heading north or south. If such fish die in large numbers in this crucial area, populations of affected fish species up and down the coast could eventually shrink.

Early in our research, as we established that *Pfiesteria* is highly lethal to fish, we also learned that fish are not its only victims; people can also be affected. Other toxic dinoflagellates generally hurt people by poisoning seafood. But studies by David P. Green of N.C.S.U. and his co-workers found little evidence that *Pfiesteria* toxins accumulate in fish, a sign that seafood harvested from *Pfiesteria*-contaminated waters probably does not serve as a "middleman" in harming human beings. Instead the exposure route is more direct: people can become dangerously ill after getting toxin-laden water on their skin or after breathing the air over areas where fish are hurt or dying from their own encounters with toxic *Pfiesteria*.

An Unwelcome Surprise

We learned about this last effect on people the hard way. When we first began our investigations, we followed established safety procedures for working with toxic dinoflagellates. We had been informed by specialists on other toxic dinoflagellates that in the laboratory contact with contaminated water was the only danger. We did not know that *Pfiesteria* produces an aerosolized neurological toxin that can seriously hurt people—the first dinoflagellate known to do so—and that we were inhaling it.

The symptoms were so subtle at first that we attributed them to other causes: shortness of breath that we ascribed to asthma; problems akin to allergy attacks, such as itchy or mildly burning eyes or a "catching" in the throat; and headaches and forgetfulness that we attributed to stress.

Before we realized that *Pfiesteria* can produce aerosolized toxin, 12 people from four different labs were sickened from toxic cultures. Three of us, myself included, have sustained some persistent problems we did not have before we began to study toxic *Pfiesteria*.

We now conduct our research in a specially designed biohazard III facility, using more precautions than are needed for most research with the AIDS virus. The lab is fitted with air locks, decontamination chambers and other safety features, and researchers wear full hooded respirators supplied with purified air.

Chronic Effects in the Field

Divers, fishermen and others working in contaminated waters while fish were showing signs of *Pfiesteria* poisoning have described respiratory problems, headaches, extreme mood swings, aching joints and muscles, disorientation and memory loss. Such anecdotal reports have recently been bolstered by formal clinical assessments.

In 1997, for example, three small outbreaks of *Pfiesteria* led Maryland's governor to close the affected waters in Chesapeake Bay for several weeks. Reports of strange symptoms in people who had been in the affected areas prompted the Maryland Department of Health and Mental Hygiene to organize a medical team to investigate. Among those who complained were heavily exposed fishermen—who described getting lost on a bay they had worked their entire lives or losing their sense of balance and concentration.

Doctors have difficulty diagnosing this "*Pfiesteria* syndrome" conclusively, however, because the specific toxins at fault have not yet been identified (as is the case with many toxic algae). Without that information, investigators cannot examine how the chemical acts in the human body, nor can tests be designed that definitively identify it in the blood or tissues. Fortunately, progress is being made. Peter D. R. Moeller and John S. Ramsdell of the National Ocean Service in Charleston, South Carolina, have semipurified components of *Pfiesteria* toxins that destroy fish skin and affect the nervous system in rats (which are studied as a model for humans).

Our own lingering health problems have led us to devote much attention to the possibility that *Pfiesteria* might cause chronic effects in fish that sustain nonlethal exposures. In lab experiments, we subjected fish to low concentrations of toxic *Pfiesteria* and monitored the animals for up to three weeks. The fish appeared to be drugged, and they developed skin lesions and infections. Tests revealed that white blood cell counts were 20 to 40 percent below normal levels, suggesting that *Pfiesteria* toxins may compromise the functioning of the immune system and make fish more susceptible to disease. Autopsies of fish that were affected have revealed damage to the brain, liver, pancreas and kidneys.

Weakened immunity, increased disease and periodic fish kills can all contribute to a decline in fish stocks. But other problems could seriously affect the ability of fish populations to recover. Research has shown that when toxic *Pfiesteria* is in the water, the eggs of striped bass and other commercially valuable fish fail to hatch.

The Bigger Picture

As we became increasingly concerned that *Pfiesteria* could threaten the viability of fish populations, we began to wonder whether this phenomenon was part of a broader trend. Dogma had long held that most finfish and shellfish exposed to sublethal doses of toxins from harmful algae suffer no ill effects. But could many harmful algae cause trouble that had been overlooked—perhaps by interfering with reproduction, with the survival of sensitive young fish or with resistance to disease? We also wondered whether there was evidence that these organisms could produce sustained or subtle health problems in people.

Few researchers have explored these questions or looked intently at the long-range effects of harmful algal blooms on the ecosystem as a whole. Nevertheless, a cluster of findings indicates cause for concern. These findings become especially disturbing when we note that as a group harmful algae are thriving. Some experts have pointed out that within the past 15 years, outbreaks of certain harmful algae seem to have increased in frequency, geographic range and virulence in many parts of the world.

Consider these examples. When bay scallops were exposed to small amounts of toxin from the dinoflagellate *Alexandrium tamarense*, their gut lining was eaten away, and their heart rate and breathing slowed. Other dinoflagellates produce ciguatera toxins that can accumulate in reef fish without killing them outright. The fish can grow large enough to be harvested as food for people, who then become sick. In fact, more human illness is caused by ciguatera-laden barracuda, red snapper, grouper and other tropical fish than by any other seafood poisoning. The symptoms can relapse for years, often triggered by alcohol consumption. Ciguatera toxins can also interfere with the normal function of white blood cells called T lymphocytes and thereby compromise the immune system. Recent work suggests that these toxins may take a similar toll on fish, resulting in impaired equilibrium, fungal infections and hemorrhaging.

Two types of cancer, disseminated neoplasia (similar to leukemia) and germinomas (which attack the reproductive organs), affect such shellfish as blue mussels and soft-shell clams. Studies have linked these cancers to certain dinoflagellates that produce saxitoxins, the same toxins that can cause sometimes fatal poisoning in people who eat contaminated shellfish. People who recover from acute saxitoxin poisoning may relapse with malarialike symptoms for years afterward.

Fish as Canaries

To combat the unwanted effects of harmful algae, scientists must first "know the enemy" more thoroughly. Many harmful algae are so poorly understood that even fundamental facts about their life cycles remain unknown. Scientists must also chemically characterize more of their toxins, so that improved warning systems can be developed for determining when waters are unsafe.

For many species of harmful algae, the factors that stimulate increased activity are as incompletely understood as the organisms' life cycles. Clearly, nutrient pollution has stimulated the growth of *Pfiesteria* and certain other members of the group. Some ecologists believe that nutrient overenrichment and other types of pollution have contributed to a serious general imbalance in many aquatic ecosystems. Large algal blooms and toxic outbreaks, they assert, are symptomatic of this imbalance as well as participants in its perpetuation.

This ecological breakdown may have many causes. Continuing losses of the wetlands that act as the earth's kidneys hamper the ability of waterways to cleanse themselves. Some algal blooms have coincided with El Niño events, suggesting that warming trends in global climate may stimulate the growth of these species and extend their range. These climatic changes also create flooding that washes additional nutrients and other pollution into rivers and estuaries. Further, inadequate environmental regulations are providing too little protection for our waters at a time when nearly two-thirds of Americans live within 50 miles of a coastline.

As we pulled anchor during the October 1995 fish kill, many thoughts were in my mind. I was keenly aware that *Pfiesteria* is but one type of harmful microorganism that can disrupt both fish resources and human health. Ultimately, water quality, human health and fish health are strongly linked. We must also become more proactive in addressing the state of our waterways, instead of reacting to each fish kill as if it were a limited, isolated crisis. In protecting vulnerable fish, the health we spare may also be our own.

How Killer Cells Kill

John Ding-E Young and Zanvil A. Cohn

The immune system is commonly likened to an army and its various cells to soldiers. The analogy is nowhere more appropriate than in the case of the cells called killer cells. Their primary duty is to seek out and destroy the body's own cells when they go wrong: to kill tumor cells and cells that have been infected by viruses (and perhaps by other foreign agents). For some years it has been clear that killer cells do their job with great efficiency, first seeking out a miscreant target cell, then binding tightly to it and finally doing something to it that causes its death, while at the same time sparing innocent bystander cells. But what exactly do they do? Just how do the killer cells kill?

Having bound to an appropriate victim, the killer cell takes aim at the target's surface and shoots it full of holes. More specifically, it fires molecules of a lethal protein. The molecules bore into the target cell's surface membrane and form porelike channels. The target cell leaks, and soon it dies.

Work in a number of laboratories, including our own at Rockefeller University, has shown that the pore-forming protein is part of the armamentarium of the two kinds of killer cell, the cytotoxic *T* cell and the so-called natural killer cell. We have found a protein with a similar function in another immune cell, the eosinophil. Moreover, the same protein, or one very like it, appears to be responsible for attacks on human cells by an amoeba that causes severe dysentery.

We begin to think pore-forming proteins may be a major weapon in a wide range of cell-mediated killing. Knowing more about the process should have important medical payoffs. It may eventually be possible, for example, to treat amebic dysentery and some other parasitic, fungal and bacterial diseases by blocking a pore-forming protein. Finding a way to enhance the pore-forming process in immune-system cells might be even more important: in principle it should be helpful in the treatment of both cancer and such intractable viral diseases as AIDS.

Once either the killer *T* cell or the natural killer cell has identified its target, the killer cell binds tightly to the target cell. This close contact triggers the lethal process and also ensures that neighboring cells are not indiscriminately destroyed.

But the nature of the killing process continued to be a mystery. The first clues to the mystery came in the early 1970s from the laboratories of a number of investigators, notably Eric Martz of the University of Massachusetts at Amherst, Christopher S. Henney of the Immunex Corporation in Seattle, William R. Clark of the University of California at Los Angeles, Pierre Golstein of the Center for Immunology in Marseilles and Gideon Berke of the Weizmann Institute of Science in Israel. Their work dissected the killing process into a sequence of discrete stages.

First, it became clear, a "conjugate" is formed as the lymphocyte and its target come in close contact. Then some kind of lethal hit is delivered, injuring the target cell. The hit initiates what seems to be a programmed death of the injured target: death proceeds in a predetermined manner (provided only that calcium ions are present) even as the conjugate breaks up and the lymphocyte goes on to initiate a new killing cycle.

Henney and Martz were among the first to suggest that a lymphocyte appeared to kill by somehow damaging the plasma (outer) membrane of its target. Their proposal was based on their observation that radioactive molecules introduced into a target cell as markers leaked out rapidly when the target was damaged by a lymphocyte. The membrane became permeable only to markers of a certain maximum size, suggesting that the damage to the membrane might take the form of holes, or pores.

Robert R. Dourmashkin and Pierre Henkart of the National Cancer Institute and their colleagues examined the surface of damaged target cells, enlarged many thousands of diameters in electron micrographs, and detected ringlike structures that appeared to be holes in the membrane. Their finding was extended by Eckhard R. Podack of the New York Medical College and Gunther Dennert of the University of Southern California

School of Medicine, who studied the effect of cultured killer cells on tumor cells. They established that the surface of the target cell was pocked with holes whose internal diameter ranged from five to 20 nanometers (millionths of a millimeter).

It was still not at all clear whether the pores were actually inflicted by the lymphocytes or simply reflected a terminal stage of cell death caused by some other form of injury. The search for an answer was long hampered by the lack of a ready source of killer cells. Then Steven Gillis of Immunex, Kendall A. Smith of the Dartmouth Medical School and Robert C. Gallo's group at the National Cancer Institute succeeded in maintaining mouse lymphocytes—both cytotoxic *T* cells and natural killer cells—in the laboratory. They were able to do so by identifying particular nutrients and growth factors required for the survival of these cells in culture; one key factor turned out to be the lymphokine called interleukin-2. Culturing made it possible to grow lymphocyte clones derived from known parent cells, and so to have at hand an abundant and homogeneous source of cytotoxic *T* cells and natural killer cells with known characteristics. The way was cleared for detailed analysis of such cells with the tools of cell biology and biochemistry.

An obvious feature of the lymphocytes clearly called for investigation. Micrographs had revealed numerous small, dark organelles (subcellular elements) in the cytoplasm. They appeared to be storage granules, which are common in secretory cells. Such granules provide an efficient means of accumulating and packaging a substance synthesized by the cell so that large quantities of it can be released quickly at the right time. The release is accomplished by exocytosis: the granules move to the periphery of the cell, fuse with the plasma membrane and disgorge their contents.

Several investigators had seen that early in the cell-killing process the granules in a killer lymphocyte become concentrated in the part of the cell that is in close contact with the target. Then Abraham Kupfer and S. Jonathan Singer of the University of California at San Diego and Dennert noted that the granule-packaging Golgi apparatus seems to be directed toward that contact region and that a number of proteins of the cell's cytoskeleton (its fibrous internal framework) "reorient" toward the target cell soon after contact is established, apparently providing the motive force that redirects the Golgi and the granules.

The reorientation of the cytoskeleton and the movement of the Golgi stacks and granules take place only when and where the lymphocyte binds to an appropriate target. Roger Y. Tsien of the University of California at Berkeley showed that the binding induces an explosive increase of calcium ions in the cell; the increase triggers exocytosis. John H. Yanelli and Victor H. Engelhard of the University of Virginia have made cinemicrographs showing granules reorienting in the cytoplasm and then fusing with the plasma membrane.

All in all, the evidence suggested that contact with an appropriate target causes the killer cell to aim its secretory apparatus at the target and fire a lethal agent contained in its granules. It was necessary first to establish that the granules are in fact the shell casings and then to identify the projectile itself.

The first task was to isolate the granules and see if they alone could kill. That was done independently in 1984 by Henkart and Podack and by our group. We exploited the various techniques of subcellular fractionation, the objective of which is to break up a cell into its components and find what component contains a particular enzyme or carries out a particular function. Killer lymphocytes are broken into bits and pieces by subjecting them to pressure in nitrogen gas. The cellular debris is layered onto a density gradient of inert particles and then spun in a high-speed centrifuge. The various organelles come to rest, depending on their density, in discrete bands. We examined each of the fractions in the electron microscope and assayed them for enzymatic activity and ability to kill cells.

One fraction, which electron micrography showed consisted almost entirely of granules, was enriched in certain enzyme activities and was a potent killer: when the isolated granules were mixed with red blood cells or tumor cells in a medium containing calcium ions, the cells died within minutes. Electron micrographs revealed that the cells' surface carried ringlike lesions virtually indistinguishable from those produced by intact killer cells. And so the granules were shown indeed to contain the killer cells' lethal secretion.

The secreted agent itself was soon identified. In 1985, collaborating with Podack and Hans Hengartner of University Hospital in Zurich, we found a protein that all by itself (in the presence of calcium) reproduces the observed membrane lesions and the killing inflicted by intact killer cells and by granules. We isolated and purified the protein by passing extracts of granules through chromatographic columns that separated proteins on the basis of electric charge and of molecular mass, and screening the separated proteins for their efficiency in lysing red blood cells: breaking down their membrane and causing them to burst. The killer protein was isolated independently by Danielle Masson and Jürg Tschopp of the University of Lausanne.

So far only this one pore-forming protein, which is often called perforin (because it perforates membranes), has been found in the granules of cytotoxic *T* lymphocytes or natural killer cells. Its molecular mass is 70,000 daltons. Once cells are exposed to perforin in the presence of calcium ions they are lysed within a few minutes. On the other hand, if calcium ions are added to perforin before it makes contact with cells, the protein's killing activity is abolished. The effect sounds paradoxical, but actually it leads to an important insight into just how perforin kills.

The 70-kilodalton molecules secreted by a killer cell insert into the target-cell membrane. There (in the presence of calcium ions) the individual molecules (monomers) polymerize, or link up with one another. The polymer they form can assume various shapes, but under optimal conditions the final product resembles a cylinder. In the electron microscope it looks like a ring when seen in cross section and like two parallel lines when seen in longitudinal section. A fully formed ring, as Podack and Dennert first noted, ranges from five to 20 nanometers in internal diameter.

For perforin to damage a target cell the calcium-mediated polymerization must take place entirely within the cell's membrane. The reason is that only the perforin monomer can insert into a membrane; if polymerization takes place in solution, in the absence of nearby membrane, the resultant polymer cannot enter the membrane and cannot kill. The protective effect is easy to understand. Any secreted perforin that spills into the extracellular space or the bloodstream, where calcium is abundant, must immediately undergo polymerization and thereby be rendered inactive, virtually eliminating the possibility of "accidental" injury to nontarget cells.

On an appropriate target cell, however, the tubular pores shaped by polymerization bring about rapid, measurable changes in the target cell. The plasma membrane of a living cell retains proteins and other large molecules within the cell (except when they are to be secreted) and segregates different ions, keeping some inside and some outside the cell. The segregation of positive and negative ions establishes a transmembrane electric potential. When the membrane leaks, ions and water tend to flow down their electrochemical gradients toward equilibration, and so there is a drop in membrane potential. If the holes in the membrane are of limited size, there is an additional effect, known as the Donnan effect. The large molecules inside the membrane cannot pass through it; water and salts from the extracellular fluid pour through the membrane toward the side that has the proteins—into the cell. The cell swells, and eventually it bursts.

By impaling target cells with microelectrodes, we were able to measure a significant drop in membrane potential soon after perforin was administered. With sensitive electronic equipment we were even able to measure the ion current flowing through individual pores. Our results were consistent with the predictions of the Donnan effect. The measurements showed, moreover, that the perforin holes are stable structures that persist as open pathways in the membrane. By determining just which ions and small molecules got through the damaged membrane, we were able to estimate that the functional internal diameter of the pores ranges from one to 10 nanometers (as opposed to the observed diameter in micrographs of five to 20 nanometers).

But what keeps the killer cell from killing itself? It cannot be the close-contact requirement, since the killer cell's membrane is continually exposed directly to secreted perforin. Chau-Ching Liu, a graduate student in our laboratory, and one of us (Young) collaborated with Clark recently to show that even purified perforin fails to kill either cytotoxic T cells or natural killer cells. The self-protective mechanism is not known, but we have a hypothesis. We think the killer-cell membrane may incorporate a special protein, which we call protectin, that is very similar to perforin. The close homology would promote a kind of "faulty polymerization": the protectin would rapidly combine with any perforin monomer that gets to the killer-cell membrane, thereby preventing either the insertion of perforin into the membrane or the normal perforin-to-perforin polymerization that would form a pore. We are now engaged in an intensive search for our hypothetical protein.

All the recent studies we have described here were done with mouse-lymphocyte cultures maintained in the laboratory. It was conceivable that perforin was just a laboratory curiosity and not truly the lymphocytes' weapon in vivo. Collaborating with Bice Perussia of the Wistar Institute in Philadelphia and Liu, we looked for perforin in lymphocytes freshly obtained from human blood. We could find none. When the lymphocytes were stimulated with interleukin-2, however, they proliferated in vitro and began to synthesize perforin. We found this to be the case for both cytotoxic T cells and natural killer cells; similar results have been reported by Leora S. Zalman and Hans J. Muller-Eberhard of the Research Institute of Scripps Clinic and their colleagues. The in vitro effect of interleukin-2 presumably reflects its effect in the body, where it is produced by helper T cells and promotes a range of immune responses.

Indeed, the effect we noted in the laboratory may explain an apparent clinical effect of interleukin-2 first reported in 1984 by Steven A. Rosenberg of the National Cancer Institute. He devised a novel therapy for certain intractable cancers in which lymphocytes extracted from the blood of a patient are stimulated with interleukin-2 outside the body and then infused back into the patient.

The lymphocytes are presumably thereby activated to kill more efficiently; some tumor regression has been observed in some patients. The procedure is highly toxic and therefore still in an experimental stage, but a better understanding of how to induce the expression of perforin by killer lymphocytes will certainly have a role in designing immunotherapies for cancer.

Predatory Fungi

Joseph J. Maio

Unless man succeeds in duplicating the process of photosynthesis, it appears that animals will always have to feed upon plants. But the plant world exacts its retribution. A number of plants have turned the tables on the animal kingdom, reversing the roles of predator and prey. These are the plant carnivores—plants that trap and consume living animals. Most famous are the pitcher plant, with its reservoir of digestive fluid in which to drown hapless insects; the sundew, with its flypaper-like leaves; and the Venus's flytrap, with its snapping jaws. But there are other carnivorous plants of larger significance in the balance of nature. We ought to know them better because they are to be found in great profusion and variety in any pinch of forest soil or garden compost. They are microscopic in size, but just as deadly to their animal prey as the sundew or Venus's flytrap.

These tiny predators are members of the large group of fungi we call molds. They grow in richly branching networks of filaments visible to the naked eye as hairy or velvety mats. Molds do not engage in photosynthesis. Like most bacteria, they lack chlorophyll and so must derive their food from other plants and from animals. Molds have long been familiar as scavengers of dead organisms, promoters of the process of decay. It was not until 1888 that a German mycologist, named Friedrich Wilhelm Zopf, beheld molds in the act of trapping and killing live animals—in this case the larvae of a tiny worm, the wheat-cockle nematode.

The nematodes (eelworms, hookworms and their like) are not the only prey of these animal-eating plants. Their victims run the gamut from the comparatively formidable nematodes down to small crustaceans, rotifers and the lowly amoeba. Charles Drechsler of the U.S. Department of Agriculture, a student of the subject for some 25 years, has identified a large number of carnivorous molds and matched them to their prey. Many are adapted to killing only one species of animal, and some are equipped with traps and snares which are marvels of genetic resourcefulness. How they evolved their predatory habits and organs remains an evolutionary mystery. These molds belong to quite different species and have in common only their behavior and some similarities of trapping technique. They present a challenging subject for investigation which may throw light on some fundamental questions in biology and may lead also to new methods for control of a number of crop-killing nematodes.

The simplest of the molds have no special organs with which to ensnare their victims. Their filaments, however, secrete a sticky substance which holds fast any small creature that has the misfortune to come in contact with it. The mold then injects daughter filaments into the body cavity of the victim and digests its contents. Most of the animals caught in this way are rhizopods—sluggish amoebae encased in minute hard shells. Sometimes, however, the big, vigorous soil nematodes are trapped by this elementary means.

More specialized is an unusual water mold, of the genus *Sommerstorffia*, which catches rotifers, its actively swimming prey, with little sticky pegs that branch from its filaments. When a rotifer, browsing among the algae on which this mold grows, takes one of these pegs in its ciliated mouth, it finds itself impaled like a fish on a hook.

Some molds do their trapping in the spore stage. The parent mold produces staggering numbers of sticky spores. When a spore is swallowed by or sticks to a passing amoeba or nematode, it germinates in the body of its luckless host and sends forth from the shriveled corpse new filaments and new spores to intercept other victims.

The most remarkable of all killer molds are found among the so-called *Fungi Imperfecti*, or completely asexual fungi. The advanced specialization of these molds is particularly interesting because they are not killers by obligation but can live quite well on decaying organic matter when nematodes, their animal victims, are not available. If nematodes are present, these molds immediately develop highly specialized structures which re-adapt them to a carnivorous way of life. They will do so even if they are merely wetted with water in which nematodes have lived.

One of these molds is *Arthrobotrys oligospora*, the nematode-catching fungus that was first studied by Zopf. When nematodes are available, it develops networks of loops, fused together to form an elaborate nematode trap. An extremely sticky fluid secreted by the mold seems to play an important role in capturing the nematode, which need not even enter the network in order to be held fast. The fluid is so sticky that one-point contact with the network frequently is enough to doom the nematode. In its frenzied struggles to escape, the worm only becomes further entangled in the loops, and finally, after a few hours of exertion, weakens and dies. The destruction these molds can cause in a laboratory culture of nematodes is appalling.

Jean Comandon and Pierre de Fonbrune have made motion pictures that show that the fungus's secretion of this adhesive substance is accompanied by intense activity in its cells. Material in the cytoplasm of the cell streams toward the point of contact with the worm. The mold may be bringing up reserves of adhesive and digestive enzymes to subdue the nematode; it may also secrete a narcotic or an intoxicant to speed the process.

Even more artfully contrived are the "rabbit snares" employed by some molds. First fully described by Charles Drechsler, these are rings of filament that are attached by short branches to the main filaments, hundreds of them growing on one mold plant. The rings are always formed by three cells and have an inside diameter just about equal to the thickness of a nematode. When a nematode, in its blind wanderings through the soil, has the ill luck to stick its head into one of these rings, the three cells suddenly inflate like a pneumatic tire, gripping it in a stranglehold from which there is no escape.

The rings respond almost instantaneously to the presence of a nematode; in less than one-tenth of a second the three cells expand to two or three times their former volume, obliterating the opening of the ring. It is difficult to understand how the delicate filaments can hold the powerfully thrashing worm in so unyielding a grip. Occasionally a muscular worm does escape by breaking the ring off its stalk. But this victory only postpones the inevitable. The ring hangs on like a deadly collar and ultimately generates filaments that invade the worm, kill it and consume it.

We are not yet sure what cellular mechanisms activate these deadly nooses. We know that in the case of the constricting ring of one mold the activating stimulus is the sliding touch of the nematode as it enters the ring. A nematode that touches the outer surface of the ring will not trigger the mechanism. But if the worm passes inside the ring, its doom is certain. This mold, then, exhibits a sharply localized "paratonic" or touch response like that of the Venus's-flytrap.

Perhaps the inflation of the cells is caused by a change in osmotic pressure, resulting in an intake of water either from the environment or from neighboring cells. Or perhaps it results from changes in the colloidal structure of the cell protoplasm. The constricting rings of one species react to acetylcholine—the substance associated with the transmission of impulses across synapses in the animal nervous system!

Some species of molds prey upon root eelworms that infest cereal crops, potatoes and pineapples. This has inspired experiments to use these fungi to control the pests. In one early experiment, conducted in Hawaii by M. B. Linford of the University of Illinois and his associates, a mulch of chopped pineapple tops was added to soil known to harbor tops was added to soil know to harbor the pineapple root-knot eelworm. This mulch produced an increase in the numbers of harmless, free-wandering nematodes which thrive in rich soil. The presence of these decay nematodes stimulated molds in the soil to develop nematode traps, which caught the eelworms as well as the harmless species. Plants protected by stimulated molds showed slight damage compared to the eelworm-ravaged control plants.

Investigators in France reported an experiment which suggests that molds may be used to control nematode parasites of animals as well as those of plants. Two sheep pens were heavily infested with larvae of a hookworm, closely related to the hookworms of man, which causes severe pulmonary and intestinal damage to sheep. One of the pens was sprinkled with the spores of three molds that employ snares or sticky nets to trap nematodes. Healthy lambs were placed in both pens. After 35 days of exposure the lambs in the pen inoculated with the molds were found free of infection, while those in the control pen showed signs of infestation with the worm.

The carnivorous molds offer many possibilities for future investigation.

One subject that needs to be explored is their role in the complex biology of the soil. We would also like to know more about the physiological mechanism that underlies the extraordinary behavior of the nematode "snares." The results of experiments on mold control of nematodes are already encouraging. They suggest that one day these peculiar little plants may perform an even more important role in agriculture than they played in nature, silently and unobtrusively, throughout the millennia before their discovery.

Extremophiles

Michael T. Madigan and Barry L. Marrs

Imagine diving into a refreshingly cool swimming pool. Now, think instead of plowing into water that is boiling or near freezing. Or consider jumping into vinegar, household ammonia or concentrated brine. The leap would be disastrous for a person. Yet many microorganisms make their home in such forbidding environments. These microbes are called extremophiles because they thrive under conditions that, from the human vantage, are clearly extreme. Amazingly, the organisms do not merely tolerate their lot; they do best in their punishing habitats and, in many cases, require one or more extremes in order to reproduce at all.

Some extremophiles have been known for more than 40 years. But the search for them has intensified recently, as scientists have recognized that places once assumed to be sterile abound with microbial life. The hunt has also been fueled in the past several years by industry's realization that the "survival kits" possessed by extremophiles can potentially serve in an array of applications.

Of particular interest are the enzymes (biological catalysts) that help extremophiles to function in brutal circumstances. Like synthetic catalysts, enzymes, which are proteins, speed up chemical reactions without being altered themselves. Yet standard enzymes stop working when exposed to heat or other extremes, and so manufacturers that rely on them must often take special steps to protect the proteins during reactions or storage. By remaining active when other enzymes would fail, enzymes from extremophiles—dubbed "extremozymes"—can potentially eliminate the need for those added steps, thereby increasing efficiency and reducing costs. They can also form the basis of entirely new enzyme-based processes.

Although only a few extremozymes have made their way into use thus far, others are sure to follow. As is true of standard enzymes, transforming a newly isolated extremozyme into a viable product for industry can take several years.

Studies of extremophiles have also helped redraw the evolutionary tree of life. At one time, dogma held that living creatures could be grouped into two basic domains: bacteria, whose simple cells lack a nucleus, and eukarya, whose cells are more complex. The new work lends strong support to the once heretical proposal that yet a third group, the archaea, exists. Anatomically, archaeans lack a nucleus and closely resemble bacteria in other ways. And certain archaeal genes have similar counterparts in bacteria, a sign that the two groups function similarly in some ways. But archaeans also possess genes otherwise found only in eukarya, and a large fraction of archaeal genes appear to be unique. These unshared genes establish archaea's separate identity. They may also provide new clues to the evolution of early life on the earth.

Some Need It Hot

Heat-loving microbes, or thermophiles, are among the best studied of the extremophiles. Thermophiles reproduce, or grow, readily in temperatures greater than 45 degrees Celsius (113 degrees Fahrenheit), and some of them, referred to as hyperthermophiles, favor temperatures above 80 degrees C (176 degrees F). Some hyperthermophiles even thrive in environments hotter than 100 degrees C (212 degrees F), the boiling point of water at sea level. In comparison, most garden-variety bacteria grow fastest in temperatures between 25 and 40 degrees C (77 and 104 degrees F). Further, no multicellular animals or plants have been found to tolerate temperatures above about 50 degrees C (122 degrees F), and no microbial eukarya yet discovered can tolerate long-term exposure to temperatures higher than about 60 degrees C (140 degrees F).

Thermophiles that are content at temperatures up to 60 degrees C have been known for a long time, but true extremophiles—those able to flourish in greater heat—were first discovered only about 30 years ago. Thomas D. Brock, and his colleagues uncovered the earliest specimens during a long-term study of microbial life in hot springs and other waters of Yellowstone National Park in Wyoming.

The investigators found, to their astonishment, that even the hottest springs supported life. In the late 1960s they identified the first extremophile capable of growth at temperatures greater than 70 degrees C. It was a bacterium, now called *Thermus aquaticus*, that would later make possible the widespread use of a revolutionary technology—the polymerase chain reaction (PCR). About this same time, the team found the first hyper-thermophile in an extremely hot and acidic spring. This organism, the archaean *Sulfolobus acidocaldarius*, grows prolifically at temperatures as high as 85 degrees C. They also showed that microbes can be present in boiling water.

Brock concluded from the collective studies that bacteria can function at higher temperatures than eukarya, and he predicted that microorganisms would likely be found wherever liquid water existed. Other work, including research that since the late 1970s has taken scientists to more hot springs and to environments around deep-sea hydrothermal vents, has lent strong support to these ideas. Hydrothermal vents, sometimes called smokers, are essentially natural undersea rock chimneys through which erupts super-heated, mineral-rich fluid as hot as 350 degrees C.

More than 50 species of hyperthermophiles have been isolated. The most heat-resistant of these microbes, *Pyrolobus fumarii*, grows in the walls of smokers. It reproduces best in an environment of about 105 degrees C and can multiply in temperatures of up to 113 degrees C. Remarkably, it stops growing at temperatures below 90 degrees C (194 degrees F). It gets too cold! Another hyperthermophile that lives in deep-sea chimneys, the methane-producing archaean *Methanopyrus*, is now drawing much atten-tion because it lies near the root in the tree of life; analysis of its genes and activities is expected to help clarify how the world's earliest cells survived.

What is the upper temperature limit for life? Do "super-hyperther-mophiles" capable of growth at 200 or 300 degrees C exist? No one knows, although current understanding suggests the limit will be about 150 degrees C. Above this temperature, probably no life-forms could prevent dissolu-tion of the chemical bonds that maintain the integrity of DNA and other essential molecules.

Not Too Hot to Handle

Researchers interested in how the structure of a molecule influences its activity are trying to understand how molecules in heat-loving microbes and other extremophiles remain functional under conditions that destroy related molecules in organisms adapted to more temperate climes. It seems that the structural differences need not be dramatic. For instance, several heat-loving extremozymes resemble their heat-intolerant counterparts in structure but appear to contain more of the ionic bonds and other internal forces that help to stabilize all enzymes.

Whatever the reason for their greater activity in extreme conditions, enzymes derived from thermophilic microbes have begun to make impressive inroads in industry. The most spectacular example is *Taq* polymerase, which derives from *T. aquaticus* and is employed widely in PCR. Invented in the mid-1980s by Kary B. Mullis, PCR is today the basis for forensic "DNA fingerprinting." It is also used extensively in modern biological research, in medical diagnosis (such as for HIV infection) and, increasingly, in screening for genetic susceptibility to various diseases, including specific forms of cancer.

In PCR, an enzyme known as a DNA polymerase copies repeatedly a snippet of DNA, producing an enormous supply. The process requires the reaction mixture to be alternately cycled between low and high temperatures. When Mullis first invented the technique, the polymerases came from microbes that were not thermophilic and so stopped working in the hot part of the procedure. Technicians had to replenish the enzymes manually after each cycle.

To solve the problem, in the late 1980s scientists plucked. *T. aquaticus* from a clearinghouse where Brock had deposited samples roughly 20 years earlier. The investigators then isolated the microbe's DNA polymerase (*Taq* polymerase). Its high tolerance for heat led to the development of totally automated PCR technology. More recently, some users of PCR have replaced the *Taq* polymerase with *Pfu* polymerase. This enzyme, isolated from the hyperthermophile *Pyrococcus furiosus* ("flaming fireball"), works best at 100 degrees C.

Others Like It Cold, Acidic, Alkaline

Cold environments are actually more common than hot ones. The oceans, which maintain an average temperature of one to three degrees C (34 to 38 degrees F), make up over half the Earth's surface. And vast land areas of the Arctic and Antarctic are permanently frozen or are unfrozen for only a few weeks in summer. Surprisingly, the most frigid places, like the hottest, support life, this time in the form of psychrophiles (cold lovers).

James T. Staley and his colleagues have shown, for example, that microbial communities populate Antarctic sea ice—ocean water that remains frozen for much of the year. These communities include photosynthetic eukarya, notably algae and diatoms, as well as a variety of bacteria. One bacterium obtained by Staley's group, *Polaromonas vacuolata*, is a prime representative of a psychrophile: its optimal temperature for growth is four degrees C, and it finds temperatures above 12 degrees C too warm for reproduction. Cold-loving organisms have started to interest manufacturers who need enzymes that work at refrigerator temperatures—such as food processors (whose products often require cold temperatures to avoid spoilage), makers of fragrances (which evaporate at high temperatures) and producers of cold-wash laundry detergents.

Among the other extremophiles now under increasing scrutiny are those that prefer highly acidic or basic conditions (acidophiles and alkaliphiles). Most natural environments on the Earth are essentially neutral, having pH values between five and nine. Acidophiles thrive in the rare habitats having a pH below five, and alkaliphiles favor habitats with a pH above nine.

Highly acidic environments can result naturally from geochemical activities (such as the production of sulfurous gases in hydrothermal vents and some hot springs) and from the metabolic activities of certain acidophiles themselves. Acidophiles are also found in the debris left over from coal mining. Interestingly, acid-loving extremophiles cannot tolerate great acidity inside their cells, where it would destroy such important molecules as DNA. They survive by keeping the acid out. But the defensive molecules that provide this protection, as well as others that come into contact with the environment, must be able to operate in extreme acidity. Indeed, extremozymes that are able to work at a pH below one—more acidic than even vinegar or stomach fluids—have been isolated from the cell wall and underlying cell membrane of some acidophiles.

Potential applications of acid-tolerant extremozymes range from catalysts for the synthesis of compounds in acidic solution to additives for animal feed, which are intended to work in the stomachs of animals. The use of enzymes in feed is already quite popular. The enzymes that are selected are ones that microbes normally secrete into the environment to break food into pieces suitable for ingestion. When added to feed, the enzymes improve the digestibility of inexpensive grains, thereby avoiding the need for more expensive food.

Alkaliphiles live in soils laden with carbonate and in so-called soda lakes, such as those found in Egypt, the Rift Valley of Africa and the western U.S. Above a pH of eight or so, certain molecules, notably those made of RNA, break down. Consequently, alkaliphiles, like acidophiles, maintain neutrality in their interior, and their extremozymes are located on or near the cell surface and in external secretions. Detergent makers in the U.S. and abroad are particularly excited by alkaliphilic enzymes.

To work effectively, detergents must be able to cope with stains from food and other sources of grease—jobs best accomplished by such enzymes as proteases (protein degraders) and lipases (grease degraders). Yet laundry detergents tend to be highly alkaline and thus destructive to standard proteases and lipases. Alkaliphilic versions of those enzymes can solve the problem. Alkaliphilic extremozymes are also poised to replace standard enzymes wielded to produce the stonewashed look in denim fabric. As if they were rocks pounding on denim, certain enzymes soften and fade fabric by degrading cellulose and releasing dyes.

A Briny Existence

The list of extremophiles does not end there. Another remarkable group— the halophiles—makes its home in intensely saline environments, especially natural salt lakes and solar salt evaporation ponds. The latter are human-made pools where seawater collects and evaporates, leaving behind dense concentrations of salt that can be harvested for such purposes as melting ice. Some saline environments are also extremely alkaline because weathering of sodium carbonate and certain other salts can release ions that produce alkalinity. Not surprisingly, microbes in those environments are adapted to both high alkalinity and high salinity.

Halophiles are able to live in salty conditions through a fascinating adaptation. Because water tends to flow from areas of high solute concentration to areas of lower concentration, a cell suspended in a very salty solution will lose water and become dehydrated unless its cytoplasm contains a higher concentration of salt (or some other solute) than its environment. Halophiles contend with this problem by producing large amounts of an internal solute or by retaining a solute extracted from outside. For instance, an archaean known as *Halobacterium salinarum* concentrates potassium chloride in its interior. As might be expected, the enzymes in its cytoplasm will function only if a high concentration of potassium chloride is present. But proteins in *H. salinarum* cell structures that are in contact with the environment require a high concentration of sodium chloride.

Investigators are exploring incorporating halophilic extremozymes into procedures used to increase the amount of crude extracted from oil wells.

To create passages through which trapped oil can flow into an active well, workers pump a mixture of viscous guar gum and sand down the well hole. Then they set off an explosive to fracture surrounding rock and to force the mixture into the newly formed crevices. The guar facilitates the sand's dispersion into the cracks, and the sand props open the crevices. Before the oil can pass through the crevices, however, the gum must be eliminated. If an enzyme that degrades guar gum is added just before the mixture is injected into the well-head, the guar retains its viscosity long enough to carry the sand into the crevices but is then broken down.

At least, that is what happens in the ideal case. But oil wells are hot and often salty places, and so ordinary enzymes often stop working prematurely. An extremozyme that functioned optimally in high heat and salt would presumably remain inactive at the relatively cool, relatively salt-free surface of the well. It would then become active gradually as it traveled down the hole, where temperature rises steadily with increasing depth. The delayed activity would provide more time for the sand mixture to spread through the oil-bearing strata, and the tolerance of heat and salt would enable the enzyme to function longer for breaking down the guar.

If the only sources of extremozymes were large-scale cultures of extremophiles, widespread industrial applications of these proteins would be impractical. Scientists rarely find large quantities of a single species of microbe in nature. A desired organism must be purified, usually by isolating single cells, and then grown in laboratory culture. For organisms with extreme lifestyles, isolation and large-scale production can prove both difficult and expensive.

Harvesting Extremozymes

Fortunately, extremozymes can be produced through recombinant DNA technology without massive culturing of the source extremophiles. Genes, which consist of DNA, specify the composition of the enzymes and other proteins made by cells; these proteins carry out most cellular activities. As long as microbial prospectors can obtain sample genes from extremophiles in nature or from small laboratory cultures, they can generally clone those genes and use them to make the corresponding proteins. That is, by using the recombinant DNA technologies, they can insert the genes into ordinary, or "domesticated," microbes, which will often use the genes to produce unlimited, pure supplies of the enzymes.

Two approaches have been exploited to identify potentially valuable extremozymes. The more traditional route requires scientists to grow at least small cultures of an extremophile obtained from an interesting environment. If the scientists are looking for, say, protein-degrading enzymes, they test to see whether extracts of the cultured cells break down selected proteins. If such activity is detected, the researchers turn to standard biochemical methods to isolate the enzymes responsible for the activity and to isolate the genes encoding the enzymes. They then must hope that the genes can be induced to give rise to their corresponding proteins in a domesticated host.

In the other approach, investigators bypass the need to grow any cultures of extremophiles. They isolate the DNA from all living things in a sample of water, soil or other material from an extreme environment. Then, using recombinant DNA technology once again, they deliver random stretches of DNA into a domesticated host—ideally one insert per host cell—without knowing the identities of the genes in those fragments. Finally, they screen the colonies that grow out, looking for evidence of activity by novel enzymes. If they find such evidence, they know that an inserted gene is responsible and that it will work in the domesticated host. This method thus avoids many bottlenecks in the traditional process. It turns up only enzymes that can be manufactured readily in tried-and-true hosts. And investigators can mine the genes for the enzymes from mixed populations of microbes without needing to culture extremophiles that might have trouble growing outside their native milieu.

Discovery of extremophiles opens new opportunities for the development of enzymes having extraordinary catalytic capabilities. Yet for any new enzyme to gain commercial acceptance, its makers will have to keep down the costs of production, for example, by ensuring that the domesticated microbes used as the extremozyme-producing factory will reliably generate large quantities of the protein. The difficulties of perfecting manufacturing techniques, and the reluctance of industries to change systems that already work reasonably well, could slow the entry of new extremozymes into commerce. It seems inevitable, however, that their many advantages will eventually prove irresistible.

Suggested Reading

Burkholder, JoAnn M. "Implications of Harmful Microalgae and Heterotrophic Dinoflagellates in Management of Sustainable Marine Fisheries," *Ecological Applications* 8, no. 1 (Supplement) (February 1998): 537–562.

Burkholder, J. M., E. J. Noga, C. W. Hobbs, and H. B. Glasgow, Jr. "New 'Phantom' Dinoflagellate Is the Causative Agent of Major Estuarine Fish Kills," *Nature* 358 (July 30, 1992): 407–410.

Carmichael, W. W. "A Status Report on Planktonic Cyanobacteria (Blue-Green Algae) and Their Toxins." *U.S. Environmental Protection Agency Report* EPA/600R-92/079 (June 1992).

Carmichael, W. W., and I. R. Falconer. "Diseases Related to Freshwater Blue-Green Algal Toxins and Control Measures." In *Algal Toxins in Seafood and Drinking Water*. I. R. Falconer, ed. Chestnut Hill: Academic Press, 1993.

Hallegraeff, Gustav M. "A Review of Harmful Algal Blooms and Their Apparent Global Increase," *Phycologia* 32, no. 2 (March 1993): 79–99.

Landsberg, Jan. "Neoplasia and Biotoxins in Bivalves: Is There a Connection?" *Journal of Shellfish Research* 15, no. 2 (June 1996): 203–230.

Epstein, Paul. "Marine Ecosystems: Emerging Diseases and Indicators of Change." In *Year of the Ocean Special Report*. Center for Health and the Global Environment, Harvard Medical School, Boston, 1998.

Packer, Lester et al., ed. "Cyanobacteria." In *Methods in Enzymology*, Vol. 167. Chestnut Hill: Academic Press, 1988.

Pearson, M. J. et al. *Toxic Blue-Green Algae: A Report by the National Rivers Authority*. London: National Rivers Authority, 1990.

Index

Fishing for hagfishes, 101
Fishing nets, giant squid in,
111
Fishing spiders, mating
among, 146
Fission-track dating, of East
Turkana geology, 10
Fitch, Frank J., 10
Flagella, of dinoflagellates,
205
Flies (Diptera), wasp
predation on, 159, 162,
163, 164
Flight, carnivorous birds and,
56
Floods, extinction via, viii
Flores island, 44, 49
Komodo dragons on, 51
Flores Sea, 44
Florets, of *Philodendron
selloum*, 173–176
Florida, 168
Flowers
inflorescences versus, 169
plant self-warming and,
172–173
Fonbrune, Pierre de, 222
Food, in symbiosis with ants,
133. *See also* Diet
Foramen magnum, of
Plesianthropus, 5
Forests, shrews in, 59–60
Forked tongue, of Komodo
dragon, 45–46
Formica ants, 117
slavery in, 126–127
wasp predation on, 161
Formica fusca, wasp
predation on, 161
Formica pergandei, slave-
making among, 126–127
Formica rubicunda, slave-
making among, 127
Formica sanguinea, slave-
making among, 127
Formica subintegra, slave-
making among, 126–127
Formica subsericea, slave-
making among, 127
Formicology, 115
Formosa, poisonous snakes
of, 66
Fossil collecting, at East
Turkana, 11–12
Fossils. *See also* Hominid
fossils

of carnivorous birds, 54,
55–56
of cyanobacteria, 195
of hagfishes, 98
of predatory wasps, 164
of varanids, 35, 49
Foxes, teeth of, 39
France, 223
Francis, George, 194
Franz Josef, Emperor, 7
Freetailed bat, 75, 76
Fruit bat, 80
Fruit-eating bats, 74
Fungi, predatory, 220–224
Fungi imperfecti, predation
by, 221–222
Funnel, of giant squid,
105–106
Funnel-eared bat, 75
Fur, of shrews, 59

Galápagos Islands
Charles Darwin at, ix
tortoises of, 32–33
Gallo, Robert C., 215
Gannon, William L., 78
Genes, for extremozymes,
232–233
Genetic screening, 228
Genghis Khan, 113
Genitalia
of giant squid, 107–108
of hagfishes, 98, 99
of Komodo dragons, 47
Geococcyx californianus,
lifestyle of, 55
Germinomas, saxitoxin and,
211
Giant marine lizards,
Cretaceous, 49
Giant monitor, 44
Giant squid, vii, 102–111,
105
anatomy of, 103–108, *104,
105*
appearance of, 102
buoyancy of, 108
discovery of, 102
eyes of, *105*, 106
feeding methods of, 109
in fishing nets, 111
habitat of, 110, 110, 111
mantle of, 104–106, *105*
maximum size of, 109–110
mouth of, *105*, 106–107
reproduction of, 107–108

size of, 102–103, 104,
109–110
strandings of, 110–111
swimming by, 105–106,
108–109, 110
taxonomy of, 102
whale predation on, 109
Giant tortoises, of Galápagos
Islands, 32–33, *33*, ix
Giant wood spider, mating by,
147
Gibbon, 6
Gibbs, James P., 32
Gigantopithecus, 6
Gili Dasami island, 49, 51
Gili Motang island, 49, 50, 51
Gillespie, Don, 46
Gillis, Steven, 215
Gills, of hagfishes, 96
Giraffe, evolution of, viii
Gladius, of giant squid, 106
Glands, of carnivorous plants,
187–190
Glasgow, Howard B., Jr., 207
Glasheen, James W., 82
Glenostictia wasp, 163
Golden digger wasp,
predation by, 161
Golgi apparatus, in killer
cells, 215–216
Golstein, Pierre, 214
Gombe National Park,
Tanzania
chimpanzees of, 21–31
predation by chimpanzees
in, 22–31
Gombe Stream Chimpanzee
Reserve, 21
Goodall, Jane van Lawick-,
chimpanzee research by,
21–22, 23
Gorham, 197
Gorilla, 6
brain size of, 4
male-female dimorphism in,
16
Göttingen, University of, 185
Gradual evolution, viii–ix
Grand Banks, giant squid in,
110
Granules, in killer cells,
215–217
Grasshoppers, wasp predation
on, 161, 164
Gray whale, as hagfish food,
95

Photo Credits